THE WORLD'S MOST MISUNDERSTOOD PREDATOR

SHARK

PAUL DE GELDER

Mudlark
An imprint of HarperCollins*Publishers*
1 London Bridge Street
London SE1 9GF

www.harpercollins.co.uk

HarperCollins*Publishers*
Macken House, 39/40 Mayor Street Upper
Dublin 1, D01 C9W8, Ireland

First published by HarperCollins*Publishers* 2022
This edition published 2023

1 3 5 7 9 10 8 6 4 2

Picture credits: p.1 (top) Image Source/Getty Images; p.1 (bottom) Cameron Spencer/Getty Images; p.2 (t) Tom Meyer/Getty Images; p.2 (b) Kåre Telnes/imagequestmarine.com; p.3 (t) Getty Images; p.3 (b) Giordano Cipriani/Getty Images; p.4 (t) by wildestanimal/Getty Images; p.4 (b) Vladimir Wrangel/Shutterstock.com; p.5 (t) Richard Robinson/Getty Images; p.5 (b) Michael Dornellas; p.6 (t) NiCK/Getty Images; p.6 (b) Marko Steffensen/Alamy Stock Photo; p.7 (t) Dr Craig O'Connell; p.7 (b) courtesy of the author; p.8 (t) Joe Romeiro; p.8 (b) Roderigo Friscione/Getty Images

A catalogue record of this book is
available from the British Library

ISBN 978-0-00-852970-3

Printed and bound in the UK using 100%
renewable electricity at CPI Group (UK) Ltd

This book is produced from independently certified FSC™ paper
to ensure responsible forest management.

For more information visit: www.harpercollins.co.uk/green

I'd like to dedicate this book to all those who work tirelessly creating films, changing legislation, fund raising and sharing their own love of sharks so that we might continue to have these wonderful creatures in our oceans for generations to come.

CONTENTS

FOREWORD

We are at war with the natural world and, in so being, we are revealing a war within ourselves. A battle of ignorance and truth, fear and understanding, corruption and morality. It may be shocking for some to learn that in the last fifty years we have seen two-thirds of all wildlife populations eradicated from the face of the earth. And beneath the waves, 90 per cent of all the large fish in our seas have gone. In fact, explorers and documentary filmmakers must now travel to the ends of the earth to find pockets of wildlife that still remain. Yet those Edens of life are ever shrinking under the threats of poaching and extraction.

It can be easy for many of us to disregard the significance of this parabolic shift from a once-abundant, to a now-scarce natural world. We can fool ourselves, quite successfully, into thinking none of this will ever touch our lives. But this naive and rather arrogant attitude is the very thing catapulting us further and further into the chaos, like the story of Icarus flying with wings made of wax and feathers ever closer to the Sun.

It's clear we have set our course on a trajectory that is in urgent need of re-evaluation, and one of the people with whom I have been fortunate enough to connect is Paul de Gelder, who is on a mission to correct our course. I was adamant about featuring Paul in my Netflix original film *Seaspiracy*, not only because he is one of the most passionate shark conservationists on the planet, but also because of his inspiring personal story of triumph and transformation in spite of the challenges life has thrown at him. It's not every day that someone on a military counter-terrorism trial is almost killed by a shark, lives to tell the tale and ends up devoting their life to promoting the protection of the very thing that almost killed them – and swimming close up with them regularly too!

Today, our psychological and digital landscapes have become saturated with narratives of fear, division and deceit. Humanity now craves answers for how we can live in harmony with each other and the rest of the natural world. The irony, of course, is that a species who has aspirationally named themselves 'humane' beings, must now look for guidance from some of the very species we have come to fear the most. Despite the bad press that sharks have received, our young species has much to learn from creatures who have inhabited the Earth before trees even existed.

Who better to take us on that journey from ignorance to truth and fear to understanding than someone who looked fear in the eye and offered compassion in return?

Ali Tabrizi
Filmmaker

INTRODUCTION

Like billions of other people around the world, to hear the word 'shark!' once filled me with terror, and why not? Think of those huge, gaping jaws filled with razor-sharp teeth. I bet that everyone reading this book has at some point worried about becoming a meal for one of these animals.

Where does that fear come from? Headlines like 'Man-eating sea monster stalks the coastline killing surfer' certainly don't help. As a young boy growing up in Australia, I spent a lot of time in the water, and movies like *Jaws* had me so scared of sharks that I'd even think about them in the swimming pools that I used to compete in (this could also be James Bond's fault, as tanks full of ravenous sharks are used by the bad guys in four – yes, *four* – of the franchise's films).

Of course, some explanation for our fear of sharks lies in the power of our imagination. I loved movies like *Jaws* and James Bond as a kid because they were stories well told, and while not every one of us will write a blockbuster, we are all the writer and director of the movies in our mind, and many of the stories that we create for ourselves are horrors. There's probably a good reason for that – perhaps the overactive imagination of our ancestors helped save them from ending up in a sabre-tooth tiger's belly – but too often in life we let the fear that we have created override fact and truth. Are some people attacked by sharks? Yes, and we'll talk about some of them, but the truth is that far more people drown in their own bathtubs than are attacked by sharks. So why is it that so many of us are afraid to dip our toe in the ocean, but we have no worries about getting into a hot bath with a glass of wine?

Australia is a country that has a reputation for deadly and dangerous creatures. A lot of people I meet around the world tell me that they'd love to visit my home country, but they're too scared of the wildlife. Believe it or not, most Australians don't ever encounter these creatures in their entire lifetime. Zoos, aquariums and reptile parks are very popular attractions with visitors and the locals in Australia. From behind thick glass, or looking down from above, anxious spectators can get a look at massive crocodiles, deadly snakes and venomous spiders.

In aquariums, you can watch the rhythmic swimming of a grey nurse shark. They may have beady yellow eyes, but grey nurse sharks have only ever bitten humans when provoked, and they've never killed one, so they're hardly the stuff of nightmares.

As for the apex predators that sit atop of the ocean's food chains, such as tiger sharks, makos and of course, the great whites, people have tried to contain these beautiful creatures in captivity, but every attempt has failed, and the animal has perished. These marvels of the underwater realm are wild and free, and the majority of humans will never see one in their lifetime – not in person, and definitely not as they're being eaten by one.

That didn't stop me being afraid of them as a kid. I'd go spearfishing with my grandfather and boogieboarding on huge swells with my brother. I knew there were sharks out there, and I constantly thought they'd want to take a bite out of me, but some things worried me more than being eaten alive, like impressing my dad. My pride was stronger than my fear, but I didn't ever conquer my terror – I simply pushed it to one side, rather than forcing myself to confront what scared me.

That was my way of dealing with most things as a kid and a young adult. I was a bit of a wayward youth, but something that I always had was a love of nature and a desire to see more of the world. To put it short, I wanted an adventure.

For a young bloke without any money, the only way for me to get that was by joining the military, and I enlisted with the Australian army in November 2000. I chose the infantry as my specialty (running around with the weight of a house on your back) and then volunteered for the paratroopers (jumping out of an airplane, crash landing, then running around with the weight of a house on your back).

My greatest memory of that time is of being deployed as a United Nations peacekeeper to a small nation in South East Asia called East Timor (also known as Timor Leste). I was the only person in my small unit of about thirty guys to take a language course in Tetum-Terik, and so, with an entire two weeks of training under my belt, I became the translator. This was one of the greatest gifts I was ever given, as I was able to experience a way of life and a people that I never would have met in my previous life as a bartender. I couldn't wait to do it again, but the army had other plans for me, and I spent the next two years in an incredibly frustrating cycle of training without deployments. In the end, I decided that it was time to move on.

I did four years as a paratrooper, where I was given a lot of opportunities to confront fear – when you parachute at night, you basically jump out into an empty, black sky and hope for the best – but I didn't ever come across any opportunities to confront my fear of sharks. Not surprisingly, you don't see too many of them when you're working in the Australian bush, and so my fear of the 'terrors of the deep' remained exactly as it had as a child. I may have grown a lot in those four years, but my fear of sharks was unchanged. It may surprise you, then, to learn that I asked to transfer to the navy to become a diver.

And not just any diver, but a clearance diver, which is an elite unit that works in small teams to conduct a variety of missions, from salvage to demolition. As you may have guessed from the name, being a navy clearance diver involves spending a lot of time in the water. I was always in the pool or the ocean as a kid, and after a break from it in the army I'd decided that I really missed it; plus these guys were based in one of the nicest parts of Sydney, and I could live on Bondi Beach, which may have factored into my decision-making too.

INTRODUCTION

The clearance divers are a secretive and highly respected branch of the armed forces, and the selection process was exactly what you'd expect from such a prestigious group. I was pushed to my limits and then further again. A lot of the training took place in Sydney Harbour, and while I was still scared of sharks, I was more scared of failure; it's no exaggeration to say that I would rather have died than give in. Sometimes you have to give yourself no other option but to succeed.

After selection and training courses for a year and a half, I served for two and a half years at clearance diving team one in Sydney. I spent a lot of my time in the water on duty, and body surfing off duty. Never in this time did I see even one of the creatures of my nightmares, but I often worried about them. What would shoot out at me from the murky depths? What was in the waves that I rode? Sharks were on my mind every time that I got into the water, but I never seemed to be on theirs.

That all changed one early morning in February 2009.

I was conducting a counter-terrorism trial that involved swimming solo around the warships in Sydney Harbour. Sharks were on my mind, but they'd never bothered me before, so I tried to rationalise that they wouldn't bother me that day. I'd spent hundreds, *thousands*, of hours in the water. Why would that day be any different?

Whack!

The 9-ft bull shark slammed into me and grabbed a huge chunk of my leg and hand in its jaws. I was like a chew toy to this predator, and when it started thrashing its jaws I was overcome by the most intense pain I could have ever imagined, and I feared that I would be torn apart and dragged into the deep. My life did actually flash before my eyes, but then a strange thing happened. It let go. I was pulled out of the water, and several surgical operations later, I woke up to find that I was missing half an arm and a leg.

Up until that day I had been terrified of sharks, but only now, after one had tried to eat me alive, would they become the primary focus of my life, and more than that.

I refused to let my injuries stop me from doing what I loved, and I went back into the ocean, and even back into Sydney Harbour. While I was definitely still scared of sharks, what scared me more was not being able to continue to do my job in the navy.

I passed every test they put in front of me, and re-qualified as a navy clearance diver, but the bottom line became clear: the navy would never allow me to rejoin my dive team or deploy on combat operations. For the first time in a long while, I began to have doubts as to what my future would hold. Who would have thought that it was the bull shark that would steer me down a new career path I could never have imagined?

As part of my recovery, I started to read more about sharks. I wanted to understand what had changed my life, and the more that I learned, the more I realised how much we – as humans – are changing theirs.

My story was followed by the Australian media, and it wasn't long before I began to receive requests to give talks to businesses and teams about overcoming adversity, and I started to consider that this could become a new career. I became absolutely fascinated by these superheroes of our oceans and their incredible abilities. When opportunity knocked, and I was given the chance to start working with sharks for TV documentaries, I jumped straight into that life and I've never looked back. I've been a presenter for *Shark Week* – an annual event on Discovery Channel – for some time now, and I like to joke that the producers are always trying to kill me (well, at least I hope I'm joking). They've thrown me from airplanes into the ocean, left me drifting in the Atlantic and the Pacific oceans for days, and had me swimming through bags of my own blood to see if it would send the sharks into a frenzy, but I'm still here and I haven't lost any more limbs since the attack in Sydney Harbour.

I wish I could say that sharks are doing just as well as me, but unfortunately, you are reading this book in a time of massive crisis for these fascinating creatures. Sharks have taken hundreds of millions of years to evolve, but the human race is killing them at a rate that will see many – if not most – species become extinct in the coming decades. We have a short time to act, or sharks will be lost to the world forever. The good news is that you as an individual *can* make a difference.

It's time to meet these denizens of the deep and find out what makes them so special, so important, and yet so endangered. Be warned, though: you may fall in love with them as much as I have.

Paul de Gelder
Los Angeles, 2022

1

THE MARVEL OF THE SEA

Shark anatomy

Sharks are winners. Not only do they exist in every sea and ocean around the world, but they are some of the oldest species on the planet, pre-dating even dinosaurs. They have survived five mass-extinction events, and despite 400 million years of evolution they remain wholly unchanged. Why? Because some things you just can't improve upon, and the anatomy of the shark is one such marvel of nature.

TEETH

Just like us, sharks lose their teeth throughout their life, but unlike humans, sharks continue to regrow them so that they're never just a set of gums.

Of course, sharks have evolved to have this trait because it's essential for their survival. They can't mash up their food by hand or with rocks like *Homo sapiens*. A shark starts dinner by biting and ripping its meal into pieces, and you can't do that without a mouthful of teeth. Prey that struggles when caught takes a toll on the hardware, and so having continuous replacements is essential for these animals to keep a mouthful of razor-sharp teeth.

Amazingly, young sharks' teeth may change shape as they get older and their food source changes. When great whites are young they mostly feed on fish, eels and squid, and for that they need sharp, pointed teeth. Then, as the great whites get bigger, their diet switches to marine mammals like whales and seals, and for this they require new tools: wide, serrated teeth for biting and tearing tough flesh and thick layers of blubber.

The sleek-looking mako shark is shaped like a missile, and for good reason: they hunt the fastest fish in the ocean, and for this makos require teeth that angle inwards so that they can snatch and hold their prey. As well as mackerel, squid and tuna, billfish such as marlin, swordfish and sailfish are the mako's prey and they can weigh up to 1,800 lb and move at speeds of up to 70 mph. They are predators in their own rights, but the mako sits above them in the food chain. When a mako gets hold of its strong prey, its backwards-facing teeth sink in like hundreds of fishing hooks, and it's usually game over. However, that doesn't mean that the billfish go to a watery grave without a fight. I once saw the vertebrae of a deceased mako that had the pointed spear of a billfish skewered through its spine. Believe it or not, that isn't what killed the mako, as the shark had lived long enough for the cartilage of the spine to heal around the wound, trapping it inside the mako until the shark later died of other causes.

As we talked about, sharks replace their teeth, so why are they able to do this when so many other species of predator cannot? Well, the reason for this is because a shark's teeth are held in the gums of its jaws, rather than being set into the bone, which is the norm for most toothy creatures, including humans. New teeth grow on the inside of the mouth, and over time, they grow and 'roll' to the outside in their new rows. The great white can have up to seven rows of exposed teeth at once and can cycle through an estimated 30,000 teeth in their lifetime – no wonder they don't get cavities.

The three largest species of shark are known as filter-feeding fish because they swim through the oceans with their wide mouths open, filtering out krill and other organisms. The whale shark, basking shark and megamouth (yes, that's its real name) are truly remarkable and enormous beasts. Their diet of plankton doesn't seem to do them any harm, as whale sharks can grow to over 60 ft in length and a scale-busting 30 tonnes in weight (which is about the same weight as a Sherman tank). This massive size helps keep the gentle creatures safe from ocean-dwelling predators, but not the ocean-going ones: *us*. These sharks are relatively slow swimmers and are very vulnerable to collisions with boats and shipping.

To reach such a size can take decades of growth. And just like counting the rings of a tree stump, you can tell the age of a shark by counting the rings on its vertebrae. Of course, this is only something you can do with animals that have passed away, and so with the live ones we just work to our best estimates based on previous data for that species. In many shark species the females are larger than the males, so sex is also something that needs taking into account. The average lifespan of a shark is about 20 to 30 years, but some can live much longer. Great whites may live up to 70 years, and it's possible that the mighty whale shark lives up to 100 years.

Yet that is nothing compared with the Greenland shark, which is the clear champion of shark lifespans. These animals are thick skinned and darkly patterned to the point where they look like swimming stones, and they can live for a staggering 500 years. It makes you realise how little time humans have been a part of the world, let alone us in our individual lives. It's truly humbling to study these magnificent beauties that share our planet.

FINS

Aside from a shark's jaws and teeth, there is another iconic feature of its anatomy which I know we are all well acquainted with.

A fin coming towards you in the water is the last thing most people want to see. This one body part can strike fear into the hearts of even the toughest person: trust me, I've seen it happen.

I remember being out on the waves on my surfboard with a few friends in the national park just south of Sydney, waiting for a wave while enjoying the sun and the ocean with my mates, when all of a sudden a large crescent-shaped fin popped up next to me and I screamed, 'SHARK!' Everybody turned to look at me, and in a micro-second their eyes had gone wide with fear, and I bet a few wetsuits got warmer. People began to paddle to shore, but I was so stunned that I didn't even move. And then another fin popped up a few feet away and water vapour was blown into the air through a blowhole. It was a dolphin! The wide eyes of my friends turned to thin slits, giving me the disappointed look that your mum gives you after you've done something stupid. In reality, I think they were all just massively relieved after having their hearts jump into their mouths.

This is the power of the fin, but they don't exist just to make me look silly. As you can imagine, a shark's fins have an incredibly important purpose.

There are the five different types of fin on a shark's body. The most predominant and well known is the dorsal fin, which rides high on the shark's back. At the back end of the shark is the tail fin, also known as the caudal fin. On each side of the shark, behind the head, are the pectoral fins. Behind these, on the underside of the shark, is the pelvic fin, and directly behind that is the anal fin. The very last fin lies opposite the anal fin on the top of the shark and is known as the rear dorsal fin. Now, each of these fins can come in a variety of sizes, shapes and different markings depending on the shark species. For example, the frilled shark doesn't have a dorsal fin at all, and the wobbegong's stretches all the way to the back of its body. The great hammerhead can have an extremely tall dorsal fin and the lemon shark has a dorsal and rear dorsal of similar size.

All of a shark's fins are there for a similar purpose – propulsion, stability and steering. The caudal fin is the driving motor for the animal, and in the case of the thresher shark, also a weapon. All of the other fins can be slightly or drastically adjusted to control the roll, pitch, depth, elevation and direction of movement through the water column; much like a pilot does with an airplane's wing and tail flaps to control its position and direction in the air. And like an airplane, the shark's fins are not designed to propel it backwards, so unless it drifts backwards in a current the shark must always move forward. This has the benefit of forcing water through the mouth and over the gills to extract vital oxygen.

A shark without fins would be like a human with no arms or legs: you wouldn't be going anywhere very quickly. And without water moving over the gills, the shark would drown. Every shark that is caught and finned will die a slow and agonising death.

THE OLFACTORY SYSTEM AND LIVER

If you look at a shark, you'll see that it has what look like nostrils at the front of its snout. These are called 'nares' and they are incredibly important to a shark because smell is one of the senses that they use to locate prey, and also to find a mate over large distances in our oceans. Many sharks have a very powerful olfactory system (the bodily structure that serves the sense of smell), and this can detect tiny molecules of scent that have been left behind by other animals as they pass through the water. In fact, this sense is so important to a shark that large parts of its brains are dedicated to this purpose alone. The white shark (also known as the great white, or white pointer) has about 14 per cent of its brain designated to its sense of smell, making it an incredibly efficient hunter and scavenger: a whale carcass can draw sharks from miles around, gathering great whites in numbers that you would rarely see anywhere else. The carcass releases scent into the ocean, and that's like ringing the dinner bell for hungry white sharks who recognise it as decomposing whale blubber – a favourite dish of theirs.

A shark's two nares act independently of each other, allowing it to use scent to zero in on its prey, even as the shark is moving. It's a level of scent detection that is almost unfathomable to us as humans.

How do we know all of this about shark nares and brains? The same way that we learn a lot of other things about them: through the necropsy of deceased animals.

If you've ever watched any kind of murder mystery on TV then you'll know what an autopsy is. A necropsy is the same, but for animals. Where an autopsy is looking for a cause of death, a necropsy can just be about expanding our knowledge of how an animal lives. Death is a part of the natural life cycle, and while we should always try and conserve the life in the oceans, we should also be realistic: nothing lives forever, and through the study of deceased animals we can learn more about the living ones and protect those of future generations that haven't even been born yet.

I wish I could tell you that the necropsy I witnessed was performed on an old shark who had reached the end of its natural life. However, that would be a lie. This poor shark was a juvenile great white who could have had another 60 years of swimming in our beautiful oceans. Instead, it had been caught in one of Australia's barbaric and outdated shark nets.

Every one of us would have much preferred that the shark was alive and swimming in its home. Instead, our scientists tried to make the best of the terrible situation and learn what they could from this poor animal.

First they opened up its brain, and I was surprised how spread out it was. Great white brains are long and Y-shaped rather than the big blob we have in our human skulls. The brain was protected by a layer of extremely thick cartilage and extremely tough to penetrate, even with tools. This shark was a juvenile, so I could only imagine how tough the cartilage on a fully developed adult must be. Although a shark doesn't have armour like an armadillo, the thick cartilage at the top of its head was just as strong as the helmet I'd worn in the military.

The next thing that surprised me was the simplicity of the shark's internal organs. Inside a human torso looks like a suitcase you packed at the end of the holiday, with far too much stuff to fit in. Not so the great white. On either side of a long stomach were two long lobes of the shark's liver, and that was about all I could see. Overall, though it was incredibly sad to think that the shark should have been alive in its home, the necropsy did allow me to become even more awed by these perfectly adapted animals.

It also allowed me to better compare them with ourselves. The human liver weighs just over 3 lb, making it the largest gland in the human body, but that is nothing compared with a great white, whose liver can account for more than a quarter of its body weight. So a fully grown 20-ft mature female weighing 4,000 lb has a liver that *weighs nearly half a tonne*. A big fatty liver like that is vital for a great white for a couple of reasons. When they make their long migrations, like they do when it's time to mate, food can be very scarce, and so on these epic voyages across the ocean the liver is used as a fuel source for the shark to draw upon. One great white – named Nicole, after the actress Nicole Kidman – was tracked travelling from South Africa to the west coast of Australia and back. That's a distance of 12,400 miles, which requires a lot of calories and energy. If you can't find it on your journey, you have to take it with you, and that's one thing that a massive liver does for Nicole and other great whites.

The shark's massive liver also assists its buoyancy, allowing it to ascend to the surface without much effort. This is especially useful during long migrations, when the shark goes into an energy saving mode that's almost like sleeping. Because the fatty liver weighs less than the water around the shark, it helps to stop the shark from descending too deep into our oceans, and instead allows it to swim on autopilot in a gradually ascending glide.

One of the downsides of having a nutrient-dense liver is that it represents a favourite food of orcas. Orcas can grow ten feet longer and weigh five times that of a fully grown great white, and they're incredibly smart marine mammals. Whether you're an elephant seal, a whale or a great white shark, if orcas are in the area then you don't want to be there too. They are hugely efficient predators, with the ability to coordinate, problem solve and manipulate their own environment to assist their hunting. Terrifyingly, orcas will even kill other animals just to play with their dead bodies and won't even eat them.

In recent times, a trail of great white corpses in South African waters was leaving scientists and locals baffled. After some serious detective work, they realised that these deaths were mainly the work of Port and Starboard, two orcas named for how their dorsal fins flipped to the left and right respectively.

At some point, Port and Starboard must have eaten a great white liver and decided that they would no longer settle for anything less than this delicacy, and they began hunting and killing one great white after another. They were the Frank and Jesse James of South African waters, but instead of holding up banks it was liver that they were after, and they killed so many great whites that the rest of these sharks got the message and left the immediate area. It was a blow not only to the animals but to the tourist operators who took people cage diving to see the great whites. I feel for them, but they will be the first to tell you that there is no controlling nature. Sometimes the animals at the top of the food chain aren't quite as high as you think, and Port and Starboard were the apex predators of those waters.

At the time of writing, the pair of orcas are still off the coast of South Africa and are estimated to be about 20 years old. Bad news for sharks, but in the animal kingdom death for one animal means life for another.

GILLS, SKELETON AND SKIN

The great white is one of my favourite species of shark, but there are so many more varieties out there. These beautiful creatures are diverse in their diet, their behaviour and their life cycles, but one thing that they all have in common is their skeleton.

Unlike other species of fish, sharks don't have any bones. Not a single one.

A shark's skeleton is actually made from cartilage. Give your ear a squeeze right now, and the bridge of your nose. That's cartilage; flexible, extremely durable, yet lighter than bone. This is another trick that makes the shark so energy efficient. The less you weigh, the less energy you need to expend to move yourself through the water. Simple and brilliant, and it adds a little more mystery to studying these creatures: one of the reasons that shark 'skeletons' are so hard to find is because the salt water of the oceans breaks down cartilage a lot faster than bone.

As well as the difference in skeleton, probably the biggest difference between humans and sharks is that they don't need to breathe the air like us, or like marine mammals such as orcas. Sharks still need oxygen, but they have a very special tool to extract it from the water: their gills.

These respiratory organs are common to all fish, and you can see them in the region of a shark's 'neck', behind the eyes. As water runs over the tiny blood vessels, they extract oxygen from the liquid, and that enters the shark's circulatory system, the blood carrying the vital gas to all parts of its body. As you can imagine, organs of tiny blood vessels are very vulnerable to damage, and so they are often protected on the outside by hard plates. As a human who is lucky enough to work with sharks, it is my responsibility to make sure that I am extremely careful not to damage this part of the beautiful animal.

There are between five to seven gill slits on each side of a shark's body, and most sharks have to swim continuously in order to force water through the gills so that they can breathe. Others like the Port Jackson, whitetip reef and nurse sharks, can lie still and suck water into and over their gills in an action called buccal pumping. If you ever see a shark lying perfectly still then it could be breathing this way, or perhaps it's a species that has a structure called a spiracle. Sharks like the wobbegong and angel have two openings behind the eyes that act as pumps, sucking water into the gills so that the shark can lie undetected and breathe even when covered in sand – for these species, remaining unseen helps keep them safe from predators, while allowing them to ambush their own prey. The spiracles are an incredible result of evolution and have served these shark species well for millions and millions of years, being far more reliable than even the best diving equipment that humans can produce.

The internal organs and skeleton of a shark are fascinating, and its skin is just as remarkable. 'Dermal denticles' cover the shark's entire body. These scales look like a carpet of overlapping teeth laid flat, and a shark is covered in millions of them to give it a protective and rugged outer later that's suited to the struggles and tribulations of life in the deep. Hundreds of millions of years of evolution have shaped these dermal denticles into the perfect form to allow a shark to slip stealthily through the water, and often at incredible speeds. As is often the case, scientists look to the natural world for inspiration, and mimic what mother nature has achieved in her creatures: swimsuits were created based on the design of dermal denticles, but they were eventually banned from competition because they gave the people wearing them an unfair advantage over their fellow competitors.

If you're ever lucky enough to stroke a shark, you'll notice that if your hand moves from nose to tail the skin feels smooth, and if you move from tail to nose it feels very rough and sharp to the touch: those are the dermal denticles, and your hand is experiencing what it's like for water to move over the shark's body.

On the other hand, let me give you an example of just how rough a shark's skin can be if you rub it the wrong way. During the filming of 'Raging Bulls' for Discovery's *Shark Week*, my team were catching bull sharks in the canals of Queensland, Australia. The canals of Australia's Gold Coast are well known as bull shark territory. Local fishermen have caught all sizes of these daunting fish, from small babies to huge monsters that could and have swallowed a large dog whole. Humans can no longer swim among the multi-million-dollar homes on the canal banks, but there is also concern among some people about swimming anywhere on the Gold Coast.

I was excited to film 'Raging Bulls' in this tourist hotspot, hoping to answer some of the questions local residents had, such as why the bulls seemed to be in such large numbers, and how they can ensure the safety of themselves and their children when entering the water. As with many shows like this, the premise was great but our time was limited. With much of the emphasis focused on finding and filming the sharks for a 40-minute documentary, it didn't leave us a lot of time to dive in deep to solve the actual problems. This is why working with actual scientists in this realm is so vitally important. Once the cameras are gone and filming is over, the scientists remain. They can use the budget boost that *Shark Week* has given them to continue their research, and hopefully the filming team get to go back next year and continue their work with them and ride on the coattails of their discoveries to bring them to a larger audience.

Of course, bull sharks and I have quite a storied history – one did eat part of me, after all – but I don't hold that against them, and I wanted to safely hold the bull alongside the boat during our study of them. To that end, I was given the job of placing the tail rope on a large bull shark that the team had just caught on the surface. To do this, I looped my right arm around the thinnest part of the shark's tail to try and hold it still, and with my left I placed the loop of rope around its tail fins. All was going well, but just as I was about to finish getting the rope in place the shark flexed its incredible strength, thrashing its body and raking its skin back across my arm. When I finished looping the rope and stood back, I saw that I had blood running down my arm – shark burn! The bull's denticles were so sharp that they had rubbed my own layers of skin away.

I've had shark burn twice now. The 8-ft bull shark gave it to me first on my right forearm, and more recently a 9-ft tiger shark gave it to me on my left while I was doing exactly the same thing: trying to secure the shark alongside a boat by holding their tail firm and prevent them spinning so they didn't get the rope wrapped around the head and gills. Take it from me, don't ever try and hug a shark, because you might end up losing a lot of skin. Denticles are an incredible form of protection for these beautiful creatures, and whale sharks even have a specific kind that covers their eyes. It's amazing just how well adapted these animals are to their environment.

There are so many wonderful shades of colour and patterns to marvel at. Whale sharks have some of the most beautiful patterning that you'll see in the natural world: a light grey underbelly, a darker shade above, and beautifully patterned with large white spots on its upper flanks. Of course, they don't look like this for us to ogle at, but for a reason: each shark species has the perfect camouflage for its environment, hunting style and, in some species, to keep it hidden from the bigger boys and girls who are higher up the food chain.

I've seen the effect of this camouflage not only in the water, but from above. I often film great whites with my drone, and they're very easy to spot when they're swimming over the white sand in clear water. I actually don't need to go far from my home in Los Angeles to spot them, as juvenile great whites especially frequent the coast of California, and a drone can often spot them with ease. Finding them in murky water, or over a reef, is a different prospect entirely. They blend in perfectly with the water around them and the coral below, and you'd never know that a 20-ft animal was moving right below your camera or boat.

Other shark species take this degree of disguise to the next level. Port Jackson, leopard, wobbegong, whale, angel and horn sharks are like the snipers of the shark world, masters of camouflage and concealment. They have developed stripes, spots, frills and swirls on their skin to help them blend into their environments. The tasselled wobbegong shark is a great example of this. Often referred to as a carpet shark because of their near flat shape, they have developed bumpy features and flowing frills that make them appear exactly like the coral in which they live, hunt and breed. They can be very hard to spot, but if you do see these sharks their tails are so pretty that it's very tempting to touch them. A lot of people have learned the hard way that this is a bad idea, as the wobbegong can quickly bend its body in half, and many unwary hands have ended up in a wobbegong's wide mouth as a result. It's never happened to me, but considering that they have rows of immensely sharp, pointed teeth, I can't imagine that it's a very pleasant experience, for the human at least. The wobbegong doesn't seem to mind, as they are very reluctant to let go.

DO SHARKS POOP?

I've held back the most important, cutting-edge question for last. The answer is: indeed they do. Shark poop is actually really useful to researchers because it gives a great insight into what the shark has been eating.

You've probably seen a goldfish poop. Despite being a fish, the shark's toilet habits are a little different. Those like the basking or whale shark – with their diets of plankton – expend waste that comes out as a dark cloud. With great whites – and their mostly mammal diet – the cloud is more on the green side, and full of nitrates. Many a cage diver has been crop dusted by white sharks.

With this basic introduction to the anatomy of the shark we've only just scratched the surface, and researchers and scientists spend their entire lives learning more and more about these beautiful beasties that come in all shapes and sizes.

2

ALL SHAPES AND SIZES

Over 500 species of shark in the world

Deep in an ocean's abyss, a mile beneath the waves, a prehistoric sea creature stirs from its lair ...

Its wide jaws are filled with rows of three-pronged, pin-thin teeth. It's long and slender like an eel, and its gills are feathered like a bearded dragon. As it moves through the water, this sea creature looks just like the mythical beasts that have been spoken about in hushed whispers for generations.

But this is no sea serpent. Growing to a maximum length of 6-ft, this inhabitant of nature's watery ravines is in fact the frilled shark, known as a living fossil because of its traits of dark brown skin and the articulation of its jaw to its cranium. You only have to take one look at it to know that this animal's near ancestors were around at the time of the earliest dinosaurs. To look at a frilled shark is to look back in time, to age hundreds of millions of years before the first humans walked the earth.

I've never seen a frilled shark in person, and that's for one very simple reason: even our best dive gear can't handle the deep ocean trenches where these creatures reside, their bodies perfectly adapted over countless millennia for life in such an uncompromising environment. One mile down, the water pressure is a staggering 2,500 psi. Breathing aside, human beings are just not adapted for that kind of weight on our organs and blood vessels. The frilled shark, once thought to be extinct, has none of these problems, and unlike most sharks it does not have the famous 'dorsal fin' that so many people associate with the creatures. Its long, cylindrical body makes it perfect for diving into the impossible depths where it hunts its prey, living mostly on a diet of squid and bony fish.

Perhaps it is because we have so little exposure to them, for obvious reasons, but the animals that come out of the ocean's immense depths always strike us as some of the most bizarre looking in the animal kingdom. This is definitely true for sharks, and I just love the look of some of these perfectly adapted creatures of the deep. When we think of sharks, it's often a picture of an animal like the great white that comes to mind, but the truth is that sharks come in all shapes and sizes.

There are flat sharks, cylindrical sharks and even triangular sharks. Some have heads like hammers, while others have noses like saws. Some sharks have huge, sharp teeth in terrifying jaws, while others have gaping wide mouths with no teeth at all. Some adult sharks will fit in your hand, while others wouldn't fit inside a school bus.

A member of the group of marine predators given the scientific name of elasmobranchs, some sharks swim in schools, while others are solitary. Some have dorsal fins and others do not. Some give birth to live young, while others lay eggs. The Greek philosopher Aristotle once (incorrectly) noted that the thresher shark hides its young in its mouth and will bite through fishing line to free itself. Because of this perceived wily intelligence, the thresher was known as the fox shark, and while Aristotle may have overestimated a shark's ability for considered thought, there's little doubt that when it comes to instinct they are second to none.

Before we take a closer look at some of the species of shark that we are lucky enough to have in our oceans, let's meet their ancestors, and see the evolutionary journey that these amazing animals have taken.

450 MILLION YEARS IN THE MAKING

Our first evidence for the existence of sharks exists in fossils that are more than half a billion years old. These fossils – in the form of scales – even pre-date the earliest known record of trees. The shark dynasty is true royalty on this planet, and the earliest shark-like teeth we have found come from 450 million years ago, which is around 380 million years before the extinction of the dinosaurs, let alone the days when they roamed the earth.

In fact, sharks have survived several mass-extinction events. Not all shark species made it through these massive biological upheavals, but enough did that they continued to evolve into the animals that we see today.

The oldest relatives of sharks were likely leaf shaped and had no bones, fins or even eyes. From here they evolved into the two groups of fish that are now in our oceans – bony fish and cartilaginous fish (which includes all sharks).

For a long time, these shark ancestors looked more like eels than what you and I would consider a shark, but the shark body type that is familiar to us today started to appear about 380 million years ago. These early days for the shark were their golden era, when these animals dominated the seas. At this time, all kinds of weird and wonderful sharks swam in the oceans and started their own evolutionary lines.

About a hundred million years after this there was another mass extinction on earth, wiping out around 96 per cent of marine life, where the previous mass extinction had only managed 75 per cent.

It was in the Jurassic period – the age of dinosaurs like the mighty Allosaurus – that sharks developed the jaws that would come to characterise them in the human mind. They continued to develop, improving in their abilities to swim and hunt, before yet another mass extinction engulfed the planet. Fossil records show that while shark species did live on, many of the larger species became extinct. This is common in mass-extinction events, as the animals at the top of the food chain are most affected by changes below them. And, because apex predators exist in smaller numbers, it's extremely hard for them to re-populate, a problem that is very much an issue in our modern day.

How is it that sharks pulled through but dinosaurs didn't? The fact that many shark species lived in deep water is one reason, because changes in global temperatures and climate affect the deep water less than they do shallow, coastal waters and of course land-based creatures. The fossil record of sharks contains more than 3,000 species. There are hundreds of species of shark currently in existence, and as we will discover they live at different depths, eat different things and come in all shapes and sizes. This biodiversity means that they have an excellent chance of avoiding natural mass-extinction events. Unfortunately, it does not save them from the one currently being carried out by human beings, but there's still time to change that. After all that sharks have been through, it would be one of the greatest crimes in the history of this planet for humanity to be the end of these animals that have been a part of our world for so very, very long.

A DIVERSE KINGDOM

With over 500 species of shark it would be impossible to give every one of them their due in this book, but let's look at a few of my favourites to give us an idea about the kind of variety in our oceans, as well as those that are most likely to be connected with our myths about these wonderful animals.

Most people have heard of nurse sharks and tiger sharks, but have you ever heard of the cow shark? They are widely considered to be the most primitive of sharks due to the fact that their skeletons and digestive systems most resemble those of their prehistoric ancestors.

There are four subspecies of cow shark, known as the bluntnose sixgill, the bigeye sixgill, the sharpnose sevengill and the broadnose sevengill. As you may have guessed, six- or sevengill refers to the number of gills on each side of the shark. Most sharks have five, so these extra ones can help to identify these animals.

There's something else that is quite remarkable about cow sharks.

Some species can give birth to over a hundred live pups.

Before you start thinking that this species is domesticated and tranquil like cattle, it's important to realise that cow sharks are predators. They can grow up to 16 ft long and often hunt in packs, meaning that they can take down fish, squid, seals and even dolphins. Human remains have also been found in cow shark stomachs, but I don't hold that against them. They can be very curious, approaching divers in a non-threatening manner. Of course, like all wild animals – particularly those with sharp teeth and strong jaws – there is a potential for danger.

Unlike the 16-ft cow shark, the lantern shark is *tiny*. In fact, it is the smallest known species of shark. It lives more than a thousand feet below the continental shelf that runs off the coast of Colombia and Venezuela. As you may have guessed from its name, the lantern shark has a special trick: it glows in the dark. This bioluminescence is due to light-emitting organs along its belly and fins, called photophores. It's thought that these glowing lights attract smaller creatures to investigate, and these are then gobbled up by the lantern shark.

It's hard for us to imagine just how dark it is at a thousand feet underwater. There is no night and day, just continuous darkness, and so another feature of the dwarf lantern is its big, bulbous eyes. You can fit one of these fully grown dwarf lantern sharks in the palm of your hand. Isn't life on our planet just amazing?

The goblin shark has always been one of my favourite species of shark, even when I was a kid and scared to be in the water. To be fair, when you look at this animal, you can understand why I was terrified: its jaws and teeth protrude beneath a massive snout which looks like the flat blade of a sword. (It's such a striking creature that they used it for the basis of Knifehead, a Kaiju monster which fights the giant robots in the blockbuster movie *Pacific Rim.*) Goblin shark sightings are incredibly rare, and studies on them are incomplete because they live at depths of more than 800 ft, with an ability to go as far down as 4,000 ft – that's almost a mile beneath the surface.

Unfortunately, most goblin shark sightings only come when they are pulled up in fishermen's nets. It must be quite a sight, albeit a tragic one. They can grow up to 13 ft in length and weigh in at about 200 lb, just under the average weight of an NFL running back. With multiple rows of extremely sharp teeth, when the goblin attacks its jaws extend out of its mouth at lightning speed to snatch its prey. Their long noses are far from ornamental and are in fact covered in sensory organs called ampullae of Lorenzini. These allow them to detect and hunt prey in the ocean's dark depths without the need to rely on their eyesight.

Have you ever heard of a stolen shark? In 2018 a horn shark was fish-napped from an aquarium in Texas. The line between madness and genius is a fine one, and the two thieves snuck it out in a stroller and dressed as a baby (a true story, would you believe). Fortunately, the stolen shark was recovered two days later and returned to its home after quite an adventure.

As you might expect, horn sharks are named for their appearance – they have two high ridges of cartilage above their eyes, and they have spines that extend from their first and second dorsal fin. They're incredibly beautiful animals, with dark dots on patterns of coffee-coloured skin. This is excellent camouflage for their hunting grounds at the bottom of the reef, allowing them to stealthily ambush their prey. Horn sharks are nocturnal predators, meaning that they hunt at night, and they use their fins like feet to creep across the reef and then strike at their unwitting dinner.

Female horn sharks reach lengths of 4 ft, but males only grow to half that size. They're no danger to humans, but they do have a powerful bite, and for good reason – a lot of the horn shark's diet is made up of molluscs and crustaceans, and without cutlery to eat these shellfish, the horn shark relies on the power of its jaws.

Pyjama sharks are named for the long black stripes that run along their silver bodies, making them seem like your grandad's old nightwear. These cute, camouflaged guys have beautiful almond-shaped eyes and little barbels above their mouths. If you're unsure what a barbel is, picture a fishy set of cat whiskers, but thicker. These barbels help the little sharks as they seek out their prey on the seafloor.

Then there's the mermaid's purse. These rectangular shapes wash up on beaches, and pyjama sharks lay two of these eggs at a time and can do this multiple times per season. These eggs have long and twisted tendrils protruding from the egg casing, perfect for attaching to marine vegetation or rocky crevices, where they will hopefully be hidden from the sight of other creatures that would want them as a snack. If they reach adulthood, pyjama sharks will grow from a pup size of 6 inches to about 35 inches as an adult. They're small and adorable, and sadly they often fall victim to fishermen's bycatch.

OCEANIC WHITETIP SHARK

Many shark species are given the names of land-based animals by the people who discover them: the tiger shark, cat shark, bull shark and leopard shark are all named for a land-going creature that shares similar looks or behaviour.

The oceanic whitetip shark is known as the wolf of the sea and was once considered the ocean's most dangerous shark. One of the world's most famous oceanographers, the French explorer Jacques Cousteau, agreed with that statement. So how did the whitetip get such a fearsome and deadly reputation?

The leading theory is that this notoriety came about following the largest shark attack in recorded history. In 1945, not long after delivering components of the atomic bomb that would be dropped on Hiroshima, the American warship USS *Indianapolis* was hit by two torpedoes from a Japanese submarine. There were 1,199 crew members on board the vessel, and as she began to sink into the waters of the Pacific Ocean there was no option for the survivors of the torpedo strike but to abandon ship. Some were able to get into life rafts, but many had to jump straight into the water to escape the blazing fires. All in all, some 900 men escaped the sinking ship, but their ordeal did not end there.

More than 200 had died when the Indy was struck, and more died of drowning and injuries in the immediate aftermath of the attack. These dead and floating bodies dispersed great clouds of blood into the water, and as we know about sharks, their sense of scent is incredible, and it wasn't long before the survivors were surrounded by animals that were feeding on the dead. When that many sharks gather in an area saturated with blood and food, they go into a feeding frenzy, and bite at things that they usually would avoid. In essence, because there's so much blood in the water, the sharks are overwhelmed by it and start biting at whatever is in front of them.

When I was attacked by a shark myself, there was so much blood in the water and in the air that you could smell it. You can only imagine how powerful that stink must be when there are *hundreds* of bleeding bodies in the water. For a time, that small corner of the Pacific Ocean must have been hell on earth. By the time that they were rescued, only 316 of the crew were left alive. That's not to say that the sharks were responsible for all of the deaths – dehydration, sunstroke and wounds from the attack would have accounted for many of the sailors – but the sharks did attack and kill, and it made for a striking story even in a war where large-scale violent death was a daily occurrence.

The story of the *Indianapolis* was one that shaped my fear of sharks as a child and young man. I'm sure we've all had that terror of floating in the ocean, with no way of knowing what's circling beneath the surface, just waiting to take a chunk out of us. I decided that the only way to get over this fear of being stranded in the ocean was to do just that: strand myself in the ocean.

Along with James Glancy – a former British special forces soldier – and a film crew, we went into the Atlantic Ocean and blew up our own boat, as part of *Shark Week*. James and I went into the water in wetsuits, wearing masks, snorkels and fins. We may have been bobbing like a pair of human corks, surrounded by water, but the human body requires fresh water, not salt. We'd be going thirsty, and hungry too, with no food allowed.

Of course, the big worry wasn't what was on our menu but whether we were on someone else's, and sure enough it wasn't long before we had company.

The wolves of the sea had found us.

Oceanic whitetips roam the open oceans, a kind of behaviour which categorises them as a pelagic species of shark. Even without humanity depleting it, food in the ocean is scarce, and so species like the whitetip are on the prowl day and night. This scarcity is also what leads to them being so inquisitive, as anything out of the ordinary needs to be investigated as a potential food source.

Like me and James.

Now we'd both done some pretty scary things in our lives, but I'd be lying if I said that I wasn't nervous.

Whitetips generally grow up to 11 ft long but have been recorded in days gone by at 13 ft, and the ocean is their home, not mine. Even wearing fins I couldn't hope to compete with these finely tuned predators, and so the key to survival was in convincing them that we were not food. When you work with sharks you get a feel for how different species behave, but neither one of us had worked with the wolves of the sea before. We'd be learning by trial and error, and in a situation like this any error could be fatal.

At first it was just James and me floating in the water, and because we weren't giving out much of a scent, the sharks were swimming around us slowly, content to just scope us out. While that was good news for keeping us out of shark jaws, it doesn't make for great television, and so our helpful crew decided it would be a good idea to start dumping some food into the mix. Predictably, this caused the sharks to become more aggressive as they chased after the chunks of fish, and what had been a calm situation suddenly felt very dangerous.

James and I positioned ourselves so that we were back to back, enabling us to look around us in a 360-degree arc. This is a lot harder than it sounds in the water, as a shark could come at us from any angle, including directly below. To say our nerves were taut is an understatement. It gave me a small glimpse into what it must have been like to be one of those stranded sailors in the war, and I understood how the whitetips had got their nickname. They certainly felt like a pack, and when I watched them bite down on the fish there was no doubt that their strong jaws could be deadly to humans. If I was to be stranded at sea for reasons other than filming a show, I'd be a lot more worried about succumbing to hunger and dehydration than falling prey to a shark. Especially now that I've had that experience.

SPINNER SHARK

The waters of the Atlantic, where I had drifted for two days with James, are home to more than fifty species of shark. One of them, the spinner shark, is an expert aerial acrobat. Spinners grow to a maximum length of just under 10 ft from nose to tail. They have black markings on the end of their fins, which understandably leads them to be confused with blacktip sharks. If you'd like to be sure of the difference between the species, just check the anal fin: if it's black, then it's not a blacktip.

These agile and effective hunters can launch themselves 20 ft in the air. (To give that some context, the legendary basketball star Michael Jordan had a vertical jump of 4 ft.) The spinner shark isn't known for its slam dunks, and instead spins its body rapidly like a tornado. This looks like a lot of fun, but the spinner isn't doing it to impress their friends – one key reason for this behaviour is to hunt. It's also suspected that they do this for predator evasion and possibly parasite removal. The spinner shark swims at high speed into a school of fish, spinning and chomping as it goes, trying to catch as much prey as it can in its needle-sharp teeth. Because the spinner attacks from below the school, its incredible momentum carries it out of the water and into the air, a breathtaking acrobatic display for anyone who's fortunate enough to witness it. Once the spinner crash-lands into the water, it gobbles up all of the fish it caught in its jaws before heading downwards to commence another attack run. The spinner is master of the perfect pirouette, and what is deadly to its dinner is a beautiful dance.

HAMMERHEAD SHARK

Most people have heard about the hammerhead shark, but did you know that there are actually nine species of them? Lots of creatures in the animal kingdom have weird and wonderful shapes to attract attention, but this shark's head is all business. Because the head is wide and spread out, it allows the highly tuned sensors within to seek out food. Have you ever seen a search party on TV, with a bunch of people walking in line to find something that could be lost in a forest? That's what the wide line of sensors in the hammerheads head is doing, searching out dinner on the reef or seafloor.

The great hammerhead is the largest of the hammerhead family, growing to about an average of 13 ft, but the largest on record was a whopping 20 ft. These predators hunt in reefs and off the coastline for cephalopods, stingrays, crustaceans and other sharks (even their own species). The great hammerhead is a cannibal, and only when these sharks mature to a certain size are they safe around their own species. Unlike a lot of other hammerheads, however, the great hammerhead tends to be a solitary swimmer.

The winghead gets its name because from above its head looks like the wide wings of an airplane. These sharks can grow up to 6 ft in length, and their head can be about 3 ft wide. A lot of times under the ocean waves it can feel like you've stepped into an alien planet, and the winghead wouldn't look out of place in a sci-fi movie.

Smalleye hammerheads are different in colour to other hammerhead species, which tend to be bluey grey. The smalleye is gold in colour, and there are a couple of theories for this. One is that the pigmentation is caused by the shrimp and catfish they eat, and the other is because the colour helps them blend into the muddy, shallow waters off South America where they live.

How can you not love a shark called the scoophead? The heads of these animals are broader and scoop shaped. They look like the front of the starship *Enterprise* was put onto a shark's body.

Scalloped hammerheads and Carolina hammerheads have more narrow heads, and the two species are almost identical. If you ever see a scalloped hammerhead, you will note that they have a thicker-shaped head with defined notches, which sets them apart from the similarly sized great hammerheads.

The bonnethead shark is a little animal, and the sides of its head are a lot closer together than other hammerheads. Its relative, the scalloped bonnethead, grows up to about 3 ft long, and these cuties can be found in the estuaries and mangrove swamps of Central America and throughout the southern waters of Texas and Florida. In fact, hammerheads are often found in brackish water, the point where the fresh water of the rivers mixes with the salt water of the sea. In August 2021 hundreds of bonnethead, blacktip, lemon and nurse sharks took refuge in a Florida canal system as a toxic red tide killed large numbers of marine wildlife off the coast. The red tide was caused by a particular kind of algae and is just another danger that sharks have to contend with. Unfortunately, hammerheads are one of the most highly sought-after sharks for the illegal trade in shark fins, and many of their species are endangered.

GANGES SHARK

Another endangered animal – critically endangered, in fact – is the Ganges shark, named after its habitat in India's river delta. The Ganges shark is not the only shark in these waters, and they can be mistaken for bull sharks, which also frequent the massive river. One big difference between the two species is their potential size. A Ganges shark will grow to a maximum of 6 ft, whereas big bull sharks have been found up to 12 ft.

Like the bull, the Ganges is short and stubby. It looks like what it is: a fast, strong, muscular predator. You might be wondering what a shark like that is doing in a fresh-water river like the Ganges. *Glyphis*, the name of the genus better known as river sharks, both live and reproduce in fresh water, while still rarely venturing into salt water. They exist in the Indo-West Pacific region, and due to fishing and human encroachment, they are all critically endangered. The Ganges is the world's largest river delta, but not even that is big enough for these sharks to escape the nets and fishing lines of mankind. Fear is another motivator for killing them, as some people believe that the Ganges shark is a man-eater. While this cannot be disproven, it is very likely that attacks attributed to the Ganges shark were actually made by bull sharks, which often swim upriver from the Bay of Bengal. When you look at a Ganges shark's jaw and teeth, they are narrow and slender: the kind that would lead you to believe the shark's diet to be fish, not other mammals. It's not like there are seal colonies on the River Ganges, and so I think it's a fair assumption that the Ganges shark would prey on the more than 250 kinds of marine fauna that exist in the delta.

Fewer than 250 of these beautiful animals are believed to be left in the world. They have a long gestation period, and small litters, meaning that their chances of population recovery are heavily stacked against them. If you're reading this a few years after I wrote it, then there's a chance that the species is already extinct.

BULL SHARK

Sharks are one of nature's success stories. They are perfectly adapted for their environment and skill, and one gold medal winner in evolution is undoubtedly the bull shark. They are the Swiss army knife of the shark world, and their ability to find food from different sources has helped them spread to all corners of the globe. You can find bull sharks in warm, shallow waters around the world, including the brackish water of estuaries, and even, on occasion, in fresh water.

The name 'bull shark' comes from the animal's shape and the fact they will often bump potential prey with their head like a bull. And like bulls they have powerful backs that grow up higher that the animal's head. These sharks are also known to have a bit of a temper, so that's another reason for their name. Bull sharks for a long time were believed to have the highest testosterone level of any animal on the planet, although this widespread myth has been shown to be untrue.

According to the Florida Museum of Natural History, there have been 117 unprovoked bull shark attacks on humans, with 25 of them proving fatal. Sounds a lot, until you understand that these attacks occurred in a period of more than a hundred years. As an average, bull sharks are recorded as making little more than one attack a year. Some estimates have the global bull shark population at 100,000, so it's fair to say these instances are not only low as an absolute number, but are also highly uncharacteristic of bull shark behaviour.

That doesn't mean that these animals should be taken lightly, of course. Make no mistake, with jaws more powerful than a great white comparative to size, and an aggressive temper, the bull shark is dangerous, but it is not the hungry man-eater it is often portrayed to be. These animals primarily eat other sharks, stingrays and fish. When food is scarce they'll broaden their diet. Some articles state that bull sharks can also slow down their digestion so that they can make meals go further when there's not an abundance of food but I could not find a study to confirm this.

In April 2013, just over four years after my attack by a bull shark, *60 Minutes Australia* aired 'Settling Scores', an episode about my recovery, which would include an adventure to Fiji to face up to bull sharks for the first time since one had nearly eaten me alive. As with every frightening challenge I faced in the military, I went into the experience trying not to think about it too much, and just let the moment unfold in front of me, confident that it would all work out.

The sheer number of sharks in the water was absolutely overwhelming; I'd never seen anything like it in my life. Six different species, and the biggest among them were the bull sharks. I steadied myself with deep breaths and watched carefully as the Fijian professionals hand fed the sharks. Gradually, any fear that was within me gave way to awe as I was able to watch these impressive creatures in their natural environment. In the final minutes of the last dive, on our final day, I fed tuna into the mouth of a bull shark, and I knew that we would be mates for life.

BASKING SHARK

While animals like the bull shark do have hot tempers and high unpredictability, other sharks glide slowly through the oceans like gentle giants.

The basking shark is one of these. With a maximum length of 35 ft, and a mouth that gapes three feet across, this incredible creature lives on a diet of tiny marine life. It doesn't hunt like many shark species. Instead, it swims steady and slow, collecting its harvest. Bull sharks and makos are sports cars of the shark world, and the basking shark is its combine harvester.

After the whale shark, the basking shark is the second-largest fish in the world. They can be found in all of our oceans, and in differing temperatures. A basking shark doesn't stick to one patch of territory but migrates around the world at a leisurely pace, scooping up its dinner as it goes.

Basking sharks cause quite a stir when they turn up on the British coastline, particularly around Cornwall. These animals swim close to the surface, and their grey colour, large size and prominent dorsal fin no doubt put the fear of *Jaws* into some people. In actual fact, basking sharks have never attacked a human, and divers have swum alongside them with absolutely no indication that the shark was bothered by their presence.

The shark gets its name because of its tendency to be close to the surface; and because it's a slow mover, this gives the appearance that it's basking under the sun. Rather than working on its tan, the chances are that the shark is close to the surface to feed on plankton. In the winter, basking sharks are not seen up top. They are sometimes found alone, but other times in schools of up to a hundred. The basking shark is vulnerable, but some countries have taken measures to protect it in their waters.

I haven't had a chance to dive with basking sharks yet, but I'd really like to. There's something mind blowing about seeing a massive creature beneath the waves. No doubt basking sharks were some of the animals that fed into the ancient myths of sea creatures. For thousands of years people have seen them as monsters, but in reality they are no danger to you unless you're plankton or a small fish.

PORBEAGLE SHARK

The basking shark is often mistaken in British waters for a great white, though there has never been a confirmed great white sighting around Britain. The closest one was in the Bay of Biscay, which is over a hundred and fifty miles from Cornwall, and on top of that, there's never actually been a single shark attack documented in Britain since records began in 1847.

That doesn't mean that all sharks in British waters eat tiny animals like the basking shark does. In fact, the porbeagle is from the order of sharks known as mackerel sharks – the same branch that includes great whites. If you were to look at a porbeagle, in many ways it does appear like a scaled-down version of these bigger animals.

That's not to say that porbeagles are small. They typically reach a little over 8 ft in length, and weigh about 300 lb. To my eye, this stout and muscular animal makes me think of it as the bulldog of the shark world, with a powerful torso, short snout and mighty jaws.

You can find porbeagles all over the North Atlantic and Mediterranean, or at least, you used to be able to.

Their numbers have been decimated by overfishing, and they are now listed as vulnerable, endangered or critically endangered, depending on which part of the world you're talking about. To put it bluntly, we've killed this species to the point where it is in danger of never recovering.

As a migratory animal, porbeagles often leave the waters where they are protected and are killed in those where they are not. In 2013 the Convention of International Trade in Endangered Species (CITES) listed the porbeagle as threatened or endangered under the Endangered Species Act.

Porbeagles are a species that has been directly targeted in some waters, and this has led to a population decline of 99.99 per cent in the Mediterranean Sea – a terrible indictment of human greed. Fishing vessels are now prohibited from fishing for large porbeagle sharks, and the species is subject to a zero total allowable catch in European Union waters, but it all seems too little, too late.

What perhaps makes this tragedy even more painful is that porbeagles are one of the few fish species that exhibit playful behaviour. Cornish fishermen and sea-goers have reported seeing these animals playing on the surface of the water, and even wrapping themselves in kelp. Porbeagles within a group have been witnessed chasing each other for sport, not unlike a pride of lions.

Though it is a formidable-looking shark, the porbeagle mostly feeds on small or medium-sized bony fish, such as mackerel and herring. I truly hope that the measures that have been put in place to protect them continue to build, as it would be an absolute tragedy to lose these animals from our seas, particularly for Britain, where it is one of the few remaining large predators in the island's ecosystem.

BLUE SHARK

The blue shark is another animal found in British waters, but where the porbeagle is stout and thick, the blue shark is long and graceful, growing up to 12 ft in length, over twice the height of your average British person.

Like many species that come to Great Britain, the blue shark is a seasonal visitor. These transatlantic animals rack up an incredible number of miles as they follow the Gulf Stream all the way from the Caribbean and back.

Their diet is mostly made up of small fish and cephalopods, and they're not afraid to go deep into the ocean to find them. In fact, the chances of you ever seeing one are extremely slim, as these animals very rarely come close to shore, and when they visit Britain they tend to stay at least ten miles offshore.

Blue sharks give birth to large litters of shark pups. It is perhaps for this reason that blue sharks have been able to weather the storm of human overfishing better than many shark species, even though they are one of the most fished sharks on the planet, and are listed as not threatened, or near threatened, depending on which organisation you talk to. What is for sure is that these animals are a truly international shark, and one which we need to work together to protect.

SPINY DOGFISH

Otherwise known as the spurdog, piked dogfish or mud shark, the spiny dogfish is a small shark belonging to the dogfish family. Although these little beauties were once one of the most common sharks in the world, their numbers are dropping dramatically due to overfishing. In European waters their numbers have been driven down so massively that it's possible they may never recover. Dinner tables across the continent have driven this demand, as dogfish has become a more popular dish. It is, however, totally unsustainable, and there are some half-hearted measures now to protect them. Left alone, the natural lifetime for these animals is over 35 years, and they have a long gestation period of two years. Given these factors, it will be an uphill struggle for their numbers to recover to even a fraction of what they once were.

I chose the spiny dogfish as an example here, but there are 118 other dogfish species. They have rougher skin than most sharks which is usually grey/brown on the upper part of their body and white below. Like a lot of sharks, dogfish hunt smaller fish and cephalopods like squid. You'll quite often see these sharks in aquariums, but beware: unlike all other sharks, dogfish have a venom which coats their dorsal fin and it is mildly toxic to humans. In the animal kingdom, defence mechanisms like this exist to convince other creatures that it's not a good idea to eat you. Unfortunately, that hasn't worked when it comes to human overfishing, and there are a critical few years ahead for these sea dogs.

SMALL SPOTTED CATSHARK

Are you a cat or a dog person? If you answered the former, you might be excited to know that this animal was once called a small-spotted dogfish before being reclassified to catshark (this was based on tooth arrangement, and shape and size of tail and fins).

This species of shark is listed as of Least Concern on the International Union for Conservation of Nature (IUCN) Red List. These little beauties have very rough skin, even for sharks. Just like their feline namesake, catsharks are predators, but instead of mice they like to eat small fish, crabs and molluscs. These animals are under 3 ft long, and with big dark eyes and big mouths, they look like a cute, little, sharky Pokémon. They are the most common shark in British seas, and it is not unknown for them to be on British menus. While they are not yet in danger of being overfished, we'll see later why eating any kind of shark is bad news for the human consumer.

GREY NURSE SHARK

The shark that you have most likely seen face to face. I have been very fortunate to dive with all kinds of sharks, but generally speaking, human beings will only come across sharks in aquariums (and of course in our overactive imaginations).

Grey nurse sharks, also known as raggy tooth or sand tigers, often form the basis of a large aquarium's shark population, and they are an incredibly important species for our understanding of sharks in general. Some species of shark, such as the great white, respond to captivity very badly – they stop eating and die. Not so grey nurse sharks, which tolerate captivity extremely well. They are also highly tolerant to being handled and tagged, making them one of the shark scientists best friends.

This doesn't mean that they don't bite. Grey nurse sharks have sharp teeth and strong jaws, and like any animal when it feels threatened, they will defend themselves. Their high tolerance for human interaction leads people to be too complacent, and when a shark is made to feel under duress it will bite. For this reason, grey nurse sharks are responsible for the fourth-largest number of shark species bites on humans, but almost all of them are provoked. We must always remember that these animals are powerful predators, and in their eyes we are a large, unknown animal. If we provoke them, we're going to get bitten – it's as simple as that. Why would we expect it to be any other way?

Grey nurse sharks grow up to about 10 ft long, but their young are about a foot, and so this species is vulnerable to being eaten by other sharks until they grow onto the bigger end of the scale.

I always used to assume that the nurse shark was given its name because it looked after its young, but like all shark species, nurse shark pups are on their own from the moment they're born. Apparently, the Old English word 'hurse' means sea-floor shark, so it's possible that 'nurse' is just a mispronunciation of that and became the accepted name over time. Nurse sharks are absolutely sea-floor animals, slowly creeping their way across the seabed and almost hoovering up any food that can fit into their mouth, where it is then chewed up by rows of teeth. Another theory is that 'nurse' comes from the ancient word 'nusse', which means catshark. Whatever the true reason, they do live up to their name when they spend their days sleeping in cosy little cuddle piles, nursing each other.

KITEFIN SHARK

At first glance, you might wonder what kind of animal you are looking at here. Long, slender, brown, and with a distinct gap between lower jaw and torso, it's no surprise that the kitefin is also known as the seal shark, as it holds a striking resemblance to that mammal. Found in many parts of the world, including the Mediterranean and the south coast of Australia, the kitefin hunts alone and preys on a wide variety of animals, including those bigger than itself. Its typical size is about three to five feet long, and it is usually found at depths of around 600 to 2,000 ft. This is an animal that hunts in the dark, and as such, it uses bioluminescence to trick its dinner into coming close. It is not the prettiest of animals, but I think it's spectacular with its huge, emerald-coloured eye.

The kitefin is no danger to humans, but the feeling isn't mutual, and these animals are hunted and exploited by mankind. Kitefins have a large fatty liver that is prized by many human industries and its meat is consumed, especially in Japan. Like many sharks, kitefins have a very slow rate of reproduction, and so they are very vulnerable to overfishing. Just as an example of how great the danger is, in the last 50 years, kitefin numbers in UK waters have dropped by 94 per cent. Apparently, 2,000 ft down isn't deep enough to escape human greed.

NINJA LANTERN SHARK

It would be a crime to write a book about sharks and not include an animal with ninja in its name. This little beauty is not to be confused with Ninja Turtles, and it's one of the tiniest shark species, with adults measuring in at under two feet long. Their eyes glow green with bioluminescence, which is thought to attract their prey. These little ninjas live at incredible depths of close to 5,000 ft, and you can just imagine them moving stealth-like through the murky waters and striking their quarry just like their namesakes.

The ninja lantern was only discovered in 2010 by shark researcher Vicky Vasquez, which really excites me. What else is in our oceans that we don't know about?

COOKIECUTTER SHARK

If you've ever watched the movie *Shark Night 3D*, you'll be familiar with these little guys. Cookiecutters can grow to around 22 inches, and they've been found at almost every ocean depth, including depths in excess of two miles.

The cookiecutter gets its name because of the shape of its mouth. Unlike most sharks, which attempt to eat an entire animal, cookiecutters cut plugs of flesh out of their prey, which vastly increases the number of species they can feed on. For a time, the US Navy couldn't figure out why small circular chunks were being taken out of their submarines' rubber-coated sonar domes. I'll give you one guess as to who the culprit was.

MEGAMOUTH SHARK

As one of the largest shark species, the megamouth is named for exactly what you think. Like the whale shark and basking shark, the megamouth feeds on zooplankton, but it does so at much greater depths than them (at least 600 ft). Like the ninja lantern, the megamouth is quite a recent discovery, and has been known to us for less than fifty years. It is an elusive animal, and fewer than a hundred of them have been sighted or caught. Growing up to 18 ft in length, these are big animals that have been found around Japan, Hawaii and California, and it is assumed that they are migratory animals.

The first specimen of this species was actually discovered entirely by accident when it became entangled in the anchor of a US warship. It sounds like something out of a movie, and it really makes you think: could there be other animals of this size out there in the deep that we don't know about? Let's hope so.

POCKET SHARK

The tiny pocket shark, discovered as recently as 2010, is so adorable and can be held in your hand. Pocket sharks actually look like miniature sperm whales. Only a few have ever been caught, so very little is known about them, but just take a moment to search for images of pocket sharks – you won't regret it.

PIG-FACED SHARK

Illustrating the fact that shark species are weird, wonderful and diverse, the official name for these animals is angular roughsharks, but they get the pig-faced name from the shape of their snout and its wide nares. Locals in the Mediterranean even insist that the shark makes a grunting noise when it's pulled from the sea. This could be caused by their wide nares sucking at air instead of water. Angular roughsharks grow up to about 3 ft long, and like most species of sharks their numbers are on the decline.

SWELL SHARK

Here's a cool party trick: the swell shark can make itself double in size. It does this by sucking in seawater, but the question is, why?

There are at least two reasons. The swell shark isn't the biggest of sharks, and that means that it's on the menu for other animals, including larger shark species. To try and look less like a snack, the swell shark puffs up its size. It's the same principle that cats use when they turn to a threat side on and arch their back. Bigger often means badder in the animal kingdom, and evolution has given this shark this neat trick.

Swell sharks like to hunt at night, meaning they need to lie low from predators in the day. Rocky crevices are a great place to hide, but holding itself in a place like that requires energy and movement that could attract predators. But it's not a problem for the swell shark. By swelling up, it can squeeze itself into the nooks and crannies and hold itself in position – a truly remarkable animal.

HELICOPRION SHARK

This species of shark has been extinct for a long time. The jaws of this animal were discovered in the Ural mountains in the late 1800s, but it wasn't until 2014 that scientists figured out what they were looking at.

So why the confusion? In the centre of this animal's lower jaw is something that looks like the disk of an angle grinder, only with teeth. In fact, this shark may well have looked like a swimming power tool. These sharks were absolutely massive, growing up to 25 ft long, and they had no upper teeth, presumably so they didn't get in the way of the circular saw of the lower jaw.

The same scientists also suggest that the helicoprion is a close shark relative rather than a shark, but I think it deserves honorary inclusion in this book for just being so bizarre.

So there we have it, not an exhaustive list of the sharks in our oceans – and hundreds more species await to be discovered – but enough to show the huge and eclectic variety of fascinating creatures out there. Now we can ask ourselves the question: how has shark societal behaviour evolved over time?

3

SHARK SOCIAL LIFE

Shark reproduction and social interaction

'Where do shark babies come from?' Now that's a really good question, because as far as sharks go there's still so much we don't know about their mating and reproduction. In the world of shark cinematography, the holy grail is capturing great white sharks mating. Nobody's done it, and the reason is simple.

No one knows where it happens.

As I mentioned earlier, great white sharks can travel *thousands* of miles. They don't stick to an isolated patch, and they're harder to track than say a pride of lions. The scale of the ocean makes the study of these creatures' lives very challenging, and a lot of the theory on great white mating comes down to educated guesswork.

One spot where scientists think that great white mating may occur is called the White Shark Café, an area in the Pacific approximately halfway between Hawaii and Baja California. The White Shark Café was named by researchers from the Monterey Bay Aquarium Research Institute and Stanford University, and it's a great white hotspot between winter and spring in the northern hemisphere. It was discovered by satellite tracking data retrieved from animals tagged off the west coast of America, and it appears as though both adult and juvenile great whites make this annual pilgrimage.

They do so at a risk: the vast ocean does not have an abundant food source like the seal colonies of California, but it seems that there is an area about 160 miles wide in the Café, known as the mesopelagic zone, where there is an abundance of food such as large fish and squid. This all-you-can-eat shark buffet is at a depth of 1,300 to 1,600 ft (0.2 to 0.3 miles) below the surface, and our finned friends make regular dives down into the deep. Thanks to the tracking data, we know that they loiter in the area for long periods of time, and one explanation for this behaviour is that it might be the great whites' elusive mating grounds.

So if we know where they go, then why has no one captured it on camera? The simple reason is that the White Shark Café is massive. In fact, it's roughly the size of the state of New Mexico, and depths can reach almost three miles down. To try and film them in such an enormous area is no small feat and would also require a lot of luck. So far no crew has had it, but I look forward to the day when we do, and we gain a little more insight into a mystery of the ocean and the beautiful creatures that inhabit it.

Some of the sharks leave the White Shark Café after a few months. Others have been recorded staying as long as 15 months (a good argument to me that there's some kind of breeding going on there). When they leave the

Café, some of the sharks travel to Hawaii, while others head to the coastal areas of Southern California and Baja California. The timing of this return trip often coincides with breeding season for elephant seals, giving the sharks a chance to gorge themself on the nutrient-rich fat and flesh of the seals.

Of course, the Pacific is huge, and there are other possibilities about where the great whites could be mating. I've done a lot of filming with the team around Guadalupe Island, which is off the west coast of Mexico. We go there to cage dive, sometimes for TV, and sometimes just because it's fun. The best months for this are September through November, and we've noticed something strange at this time that has made us wonder if it could be attributed to mating. Some of the female sharks turn up with large bite marks in the area of the head and gills. Rather than this being some kind of kinky shark foreplay, we expect that these bite marks have a practical reason.

We can make a confident, educated guess that they are related to great whites' mating because other shark species have been caught in action exhibiting the same kind of behaviour. For many species, this technique is critical in passing on their genes and blood line. Sharks don't have hands, and in their slippery and wet environment, the only tool available to a male for grabbing hold of their mating partner is their teeth.

This isn't a problem for most fish, the majority of which lay eggs which are then fertilised by a male. Sharks, however, have intercourse, and many carry their eggs internally.

Unlike a mammal, a male shark doesn't have a penis. Instead it has two sexual organs that are known as claspers. These lay flat on the underside of a shark's body, and when he has a firm grip on a female – usually by using his teeth – the male will use one of its claspers to impregnate her.

There is another peculiar shark trait that plays a role here. If you've watched *Shark Week*, or other shark-related TV shows, then you've probably seen a shark put to sleep by the team as they roll it onto its back. This is known as tonic immobility, where the mind and body of the shark seem to shut down, and the animal remains calm and gentle. As expert shark divers, we put sharks into this state by vigorously rubbing their nose when they come to take a look at us or by flipping them onto their backs. The nose rubbing in particular overstimulates their sensory organs known as the ampullae of Lorenzini which they use to detect prey, and the sharks often become tonic and cease to move. I've done this underwater many times so that we can place fin cameras on Caribbean reef sharks in the Bahamas. It is safe for all involved, and the cameras help us study more of the sharks' movements and interactions. I've never tried to put a great white into tonic, though. Something tells me that wouldn't be the smartest idea.

Of course, these animals didn't evolve to have tonic immobility just to help *Shark Week* out, and there is a convincing theory that sharks go into tonic so that they do not injure themselves during mating. A male biting into the neck with just enough force to hold on is one thing, but if the female started thrashing it could cause chunks of flesh to be ripped away, possibly leading to death.

Male sharks also go into tonic immobility, which some people might say is a reason to disprove its use for mating. But then males have nipples, and we can't breastfeed our young, so maybe we're looking at a similar kind of thing here – tonic immobility is something that both sexes share, but only one of them uses it for its purpose. On the other hand, we could be totally wrong about it having anything to do with mating at all. That's one of the great things about working with sharks and nature. We've still got so much to learn.

Some sharks lead a solitary life, and so finding a partner for mating can be a daunting task. For other species, they don't have to look far. The reef shark and hammerhead often come together in schools of thousands: anyone can find a date with those kinds of numbers. The rates at which sharks become sexually mature is also different from species to species, but generally speaking, most sharks are slow to sexually mature. Female great whites, for example, may not reach mating age until they're 33 years old, but what we usually judge their sexual maturity on is their size: a female great white of 14 ft or more is considered sexually mature. Males mature quicker and they will be about 11 ft long and 26 years old when they are ready to mate. Some smaller great whites have been documented with claspers full of sperm, but the obstacle to them mating may be that they're not big and strong enough to hold onto their partners just yet.

Our friend the Greenland shark is not thought to reproduce until it's 150 years old. That sounds like a crazy number, but when you consider that they can live upwards of 500 years, it makes a lot of sense and is comparable to the same sexual maturity/lifespan ratio that you see in a lot of sharks.

Generally speaking, across all shark species, it takes five to ten years for a shark to reach an age where they can reproduce, and when they do, their litter sizes can be just as varied.

These are more generalisations, but to give you some idea on shark litters, it's thought that bull sharks have between six to eight pups (although one particularly heavily laden mother was documented having fourteen). Sand tigers, also known as the grey nurse, may only have two pups. White sharks tend to have ten. Now, at the other end of the spectrum, tiger sharks may give birth to up to 80 young. How on earth does a mother keep track of all of these babies?

Put simply: they don't. Sharks are not maternal. Once the pup is birthed, it's on its own and at the mercy of the ocean and its inhabitants. This has led people to believe that sharks are not social creatures, but that's not the case. For a long time, great whites were thought to be lone predators, roaming the ocean in a solitary life, and only coming across other great whites at plentiful food spots like Cape Cod, or the Neptune Islands in Australia. Now, thanks to exploration of newly documented white shark hotspots and the introduction of tracking systems, what we actually see is a trend among these sharks towards some form of socialisation. It's not the affectionate relationship you would expect from a pride of lions or other mammalian hunters, but there is at least a familiarity and an acceptance between these predators.

SHARK MIRACLES

In an Italian aquarium was a female smooth-hound shark. According to articles written about the event, she had been in captivity for ten years, never encountering a male of her species. You can imagine the staff's surprise, then, when one day this lady shark popped out a pup.

Could it be that this animal had an insanely long gestation cycle? On the contrary, this was the work of parthenogenesis, a process which is known in the animal kingdom but had never been recorded with sharks. To break the Greek word down, 'partheno' means virgin, and 'genesis' means origin. This clever biological trick works when a shark creates eggs along with a few other products called 'polar bodies'. Much of our knowledge of this process is incomplete; what we do know is that polar bodies are usually reabsorbed into the body, but in the case of parthenogenesis a polar body acts like sperm and fertilises the egg. In the wild, bonnethead and zebra sharks have been observed giving birth without a preceding male interaction, leading us to believe that this process is more common in the shark world than anyone knew.

Parthenogenesis is just one part of the weird and wonderful world of shark mating. There are three other methods of shark gestation and delivery:

Oviparity

This is a nice and simple process, and just means that a shark is laying an egg. About 30 to 40 per cent of shark species fall into this category, including the bamboo, horn, wobbegong, cat and swell sharks. These examples are small species of shark, and oviparity tends to only occur in this size of animal species. The bigger ones have a different process, which we'll come to shortly.

Just like shark species, shark eggs come in different shapes and sizes. If you've ever walked down a beach and seen a strange brownish spiral that looks a bit like a piece of seaweed, then you've come across a shark egg. The egg is this shape so that the mother can fit it in secure places, such as between rocks or in underwater vegetation. This gives the egg a better chance of remaining undetected and uneaten. Some of these eggs are covered in a sticky mucus and others have tendrils. It's all designed to keep the eggs where the mother puts them. Just like a chicken egg, inside the shell is a tiny embryo that is fed by a super-nutrient-rich yolk sack. Once the yolk is totally consumed, and the now baby shark gets hungry, it will break through the egg as a fully formed shark pup. It will do this without help from its mother, because she is long gone. Once hatched from the egg, the pup is on its own and will need to find its own food while trying not to become a meal itself. These early days are the most dangerous for sharks. Eventually, some of them will have places high up the food chain, but as recently hatched pups, they're on a lot of marine menus.

If this sounds fascinating to you, then you can actually get involved with the Great Eggcase Hunt organised by the Shark Trust. You can do it anywhere in the world where there's a beach, or wherever you snorkel and scuba. The project was launched in 2003, and by collating the reports of citizen scientists around the world, the Shark Trust is able to build a picture of the state of shark breeding around the world.

Ovoviviparity

In this process, the growing foetus and egg will remain inside the mother shark until the baby shark is ready to hatch. In most ovoviviparous sharks, this part of the process tends to take about three months. Then, once the pup has exited the egg, it remains within its mother, feeding on the remainder of the yolk sack, as well as on nutrient secretions from the mother. Eventually, they will leave the mother – the process of birthing – and start their new lives alone.

The length of time it takes for a foetus to reach the optimal size for live birth differs by species. For great whites, the gestation period is 12 to 18 months. By the time that they emerge from the mother, the pups are a whopping 3.5 to 5 ft and can weigh just under 80 lb. The first great white that I ever saw in my life was about that size. It was a truly amazing experience to see it, and it gave me a lot of respect for how hard its mother must have worked to keep herself and her pups alive. The scientist that I was with at the time pulled the juvenile great white into the boat to do a workup, which is when blood and tissue samples are taken. So that the shark could breathe, we put a hose in its mouth to push water over its gills, and this not so little fella decided that he would bite down so hard on the hose that we couldn't get it back out. We tried to move him onto the rear deck of the boat to put him back in the water, but no matter what we did, he just wouldn't let go. We got so desperate we tried tickling him – fun fact, sharks are not ticklish – and in the end we managed to slide his tail into the water far enough that he recognised his natural element. That did the trick, and with a flick of his tail he was off. It was an honour to meet my first great white, and what really struck me about the animal was that he looked exactly like an adult great white, just shrunk down in size.

Things inside some mother sharks can get a little 'dark'. Often, the first pup out of the egg won't just eat its own egg, but the remainder of the eggs inside the mother, including whatever is inside them, so its brothers and sisters.

This is called 'intrauterine cannibalism'. Researchers studying sand tiger sharks noted there might be 12 pups inside the womb, but in this shark hunger games only one will emerge after consuming the rest. This tends to be the biggest, strongest shark, and these acts of cannibalism fit right into the old saying that natural selection is about the survival of the fittest. It's a shark-eat-shark world …

If a shark eats its live siblings outside of the egg it's called 'embryophagy', and sand tigers are the only species we know that do this for sure. If a shark eats unhatched siblings, and other unfertilised eggs, then this is known as 'oophagy'. It is shocking for us humans to think about this, but remember that the shark is born into a very unforgiving world. If a young shark can fall victim to its sibling, it will almost certainly be picked off by the predators that are waiting for just such a meal. It's brutal, but then that's nature: hard and unforgiving.

Some more examples of ovoviviparous sharks include basking sharks, tawny nurse sharks, all whale sharks and the entire mackerel shark order, such as the mako, white, porbeagle and salmon sharks.

Viviparity

This method of gestation and delivery involves the mother shark giving birth to live babies that have been fed by the placenta, just like you and I were. The nutrients are fed to the foetus directly from the mother's bloodstream, and as a result, there's no need for eating brothers and sisters (at least not yet).

Some of the sharks that give birth in this fashion are the hammerhead group, blue sharks, bull sharks and lemon sharks. Their litters range from two to twenty pups, and these little guys come out with the instincts to hide and start hunting as soon as they leave the mother.

DIVING WITH GREAT WHITES

After my earlier life experience of being partially eaten by a shark, it may surprise you to hear that in 2017 I found myself outside of a shark cage – willingly – and sharing the water with a group of mature great whites.

Let's rewind a little so that I can tell you how I got to that point. Andy Casagrande is an expert shark cinematographer whom I have collaborated with on several shark documentaries, and we get on really well. So, when I was told that I'd be going with him to a rocky outcrop off Western Australia known as Salisbury Island, I was ecstatic. I'd done a lot with sharks in my TV career up to this point, but I wanted to do more with the *T. rex* of the shark world, and Salisbury Island is known as a great white hotspot.

As soon as I got into the cage with Andy that day, I knew that I was about to have one of those moments that I would remember for the rest of my life. From the boat we'd spotted seven big sharks, and now we would descend 80 ft in the cage so that we could get some underwater filming – at no point did I expect to exit the cage, but Andy has a way of making things happen. As the magnificent predators swam above us, Andy opened the cage door and left the relative safety of our bars to get a better camera angle. I'd grown to like the bloke, so I decided that I'd better go after him to watch his back, and that's how I ended up in the kelp at the bottom of the sea, totally defenceless as some of nature's most perfect killers swam above me.

That dive was an awe-inspiring, life-changing moment for me, and thinking about it now I am overwhelmed with gratitude for those opportunities where I could share the water with such glorious animals. At the time I was thrilled to my bones, and over the next few days and more dives, Andy and I witnessed something that had possibly never been seen before, and had certainly not been documented: large, male great whites swimming side by side.

Previous to our encounter and filming, it was thought that sharks only ever swam alongside each other to gauge each other's size, and which one has dominance. To Andy and me, the two sharks we were watching seemed to be doing it just to hang out. Time and time again the two sharks, which were exactly the same size and with very similar markings, would come together pec fin to pec fin before swimming over the top of Andy. They'd then split up, breaking right and left into a loop before forming up and coming over for another pass. It was absolutely mind-blowing to watch something that few, if any, human beings had ever witnessed before.

So, were the sharks coming over because they liked the feeling of Andy's air bubbles tickling their pretty white bellies? Not according to Andy, who has his own theory. All of the sharks in our immediate vicinity were almost exactly the same size, and with visibly similar markings. They showed no signs of aggression at all towards each other, and Andy wondered if perhaps these were pups from the same litter. After all, they had all been in this exact spot a year earlier, and they had returned to Salisbury Island at the same time: were they coming back to the place they had been birthed?

'Why didn't you take DNA samples?' I hear you asking me, and I agree: the only way to solve this riddle would be to look into the sharks' genetics. A great white isn't going to let you put a cotton swab into the corner of its mouth, and so the only way to take the sample would be with a specialised spear pole that would take a sample of flesh containing DNA. This kind of procedure requires permission from the authorities, and alas, we were not granted it. The mystery of the isle of jaws remains unsolved, and it's something I would love to know more about one day. Having been a young man and a paratrooper, I've seen how well males can get along with each other before the females arrive, so maybe we were just witness to the calm before the storm. I really hope that one day I'll get lucky enough to go back and find out.

COMMUNITIES AND PACKS

The waters off the Bahamas are pristine, with clear blue sea above white sand making it extremely easy to both spot and film sharks. They remain a favourite location for shark documentary crews. On these trips, schools of lemon sharks very quickly turn up at the back of our boat. This isn't because they're desperate for fame and life on the silver screen, but because these clever sharks have come to associate the presence of boats with food. The tour guides promise shark viewings, and the easiest way to make good on that promise is by throwing tasty morsels into the water so that the sharks are drawn by the scent. The sharks have seen the trick long enough to know that when they hear and see the boats, it's time for dinner.

Sometimes sharks can convene around a food source from different places, bringing solitary animals together, but that's not the case for the lemon sharks. They live in close-knit groups, and the reason for this is survival. It takes a lemon shark about 12 years to reach mature age, and smaller sharks are always in danger of falling prey to other shark species such as bull and tigers. There is safety in numbers, and there are other advantages too: lemon sharks hunt together, driving prey before them like a pack of big cats, and because lemon sharks always have a mate close at hand, they don't need to make long, solitary voyages across the sea to find one.

Lemon sharks aren't just hanging out for the safety and the sex, though. They become friends with other sharks who are about their own size, communicating through body language. Lemon sharks have a large brain compared with other sharks, and scientists believe this is why the lemon shark is capable of forming such bonds and allowing higher levels of communication between these buddies than is common in most fish species. In many ways, the social life of lemon sharks is closer to that of mammals.

There is also growing evidence that these lemon sharks have their own distinct personalities, too. In a study with young lemon sharks, it was observed that some of the animals were more standoffish, while others initiated and accepted contact from other sharks. These animals that put themselves at more risk, paid for their adventurous side more often. Due to being picked off by bigger predators, the 'extroverted' lemon sharks had a lower survival rate than those observed in the study that preferred more solitude.*

* https://www.nationalgeographic.com/animals/article/sharks-form-years-long-friendships-dispelling-myths

Another shark species that lives and hunts in packs is the whitetip reef shark. This is one of the shark species that has the ability to pump water over its gills, meaning that they do not have to be constantly on the move to keep water flowing over the blood vessels in their respiratory organs. This allows them to save energy, and also hide away in the daytime hours. Whitetip reef sharks hunt at night, but some sharks hunt during the day, and whitetips could easily end up on the menu. To that end, it's not unusual to find them chilling out in a cave during the day, waiting for the waters to turn dark so that they can set out and stalk their own prey.

Studies of grey reef sharks did a lot to disprove the belief that sharks are lone wolves of the sea. This is a truly fascinating part of the animal kingdom that we still have so much to learn about. What's more, we have more in common with them than we thought for a long time. Gone are the days when sharks were thought of as unthinking, unfeeling killing machines. Just like humans and many other animals, sharks have complicated social structures. They have bonds, and family, and friendships. Somehow, that makes the knowledge of what is happening to our oceans even more painful. It's time that we stop seeing sharks as commodities that need to be harvested. The more we learn about sharks, the more we realise how incredible they truly are.

4

EAT OR BE EATEN

The shark as prey and predator

As a kid I used to dream about a shark eating me alive.

But when that nearly happened in 2009, as I encountered that bull shark in Sydney Harbour, it was not a fun experience whatsoever. That day I found myself on a different tier in the food chain to the one which I'm accustomed to. You see, we may have cars, and phones, and even space stations, but when it comes to being a lonely predator, we humans are pretty useless. We're slow. We're not as agile as our food. We're not even very strong. Our brain and our ability to work together have helped us survive and thrive as a species, but without tools we are a very lame predator.

Not the shark. Millions of years of glorious evolution have crafted the shark into the perfect predator. Their streamlined body allows them to cruise through the water like a torpedo when it's time to strike their prey. Their rough skin is so incredibly adapted to swimming that human swimmers have been banned from using imitations of it in competitions. Though some sharks have good eyesight, their sensory organs enable them to detect even the smallest movement or scent before they even see what caused it. Sharks' sense of smell is so powerful that they can home in on mates and dinner from miles away. Their teeth are like rows of the most expensive knives, perfectly adapted for killing their prey – bent back and needle-like for fish, thick and jagged for marine mammals – with powerful snapping jaws or wide gaping mouths depending on the diet of the particular species of shark.

With hundreds of species, many sharks have developed unique and extremely specific hunting techniques. Take the epaulette shark, for example. These spotted beauties are about 3 ft long and part of the long-tailed carpet shark family, found in the warm waters around Australia, Indonesia and Papua New Guinea. The epaulette's style of hunting is something that you'd never even consider for a shark.

It walks on its feet.

Except, of course, a shark doesn't have feet, so what it's actually walking on is its fins.

So why is it doing this? The epaulette shark is a coastal creature and most active at night. When the tide is out, parts of the reef are exposed, and as the water draws back it leaves small pools of water unconnected to the sea until the tide comes back in. The clever epaulette shark doesn't miss this opportunity of 'fish in a barrel' and uses its fins to get it from one oasis to the next. The epaulette shark is literally a fish out of water, and by doing this at night it avoids becoming a meal for seabirds, and instead gorges itself on the prey that is left behind in the tide pools.

You're probably wondering how this shark breathes while it's out of the water. After all, sharks and other fish need water moving over their gills. Without that, they can't extract the oxygen that is vital for survival. While it's out of the water, an epaulette shark's body switches into low power, slowing down its heart rate and respiration, and lower heart rate + lower respiration = less oxygen used. It also increases blood flow to the brain and shuts down the flow to non-essential neural functions. Because of this remarkable evolution the epaulette shark can stay out of the water for long periods – they have been recorded doing this for up to an hour! It is truly an incredible evolutionary innovation, and one which pays dividends. The epaulette shark gets an all-you-can-eat rock-pool buffet, and they will devour almost anything they can find, from worms, to eggs, to crustaceans. Eating hard-shelled animals can be tough work, but the epaulette shark has small triangular teeth to pierce and crush shells, and their gill slits expel sediment and bits of shell that they consume during the struggle to eat a live crab (creatures that don't give up without a fight).

These brilliantly adapted raiders of the reef are an egg-laying shark, producing 20 to 50 eggs a year. If the little embryos can last long enough to hatch into baby sharks, then they will take about seven years to reach sexual maturity. This means that the species is slow to reproduce, and therefore vulnerable to losses in its population. Although the epaulette shark has developed many traits, it has unfortunately not developed one that allows it to escape destruction of its habitat and numbers by humans.

HYENAS OF THE SEA

Growing up in Australia you develop a healthy respect for saltwater crocodiles. They've been on our planet since the days of the dinosaurs, and like sharks they are a perfectly adapted predator, with massive jaws powerful enough to crush bone. There are some absolutely monstrous saltwater crocs in Australia, and as their name suggests, they can live in saltwater as well as the fresh water of swamps, lakes and rivers.

I was part of a TV shoot deep in the prehistoric outback of Australia's Northern Territory. It's a harsh environment, and nothing survives there unless it's tough. At the top of the territory's food pyramid is undoubtedly the saltwater croc, and so you can only imagine my surprise when we came across someone who was happy enough to steal its dinner.

This daring heist was pulled off by a group of sharks. Maybe they were crazy, but they were definitely successful, and they managed to take food from a predator who could crush them with one bite of its massive jaws.

So how did they do it? Like hyenas, lions or wolves, these sharks were successful because they worked as a team. They'd wait until the croc had food in its jaws, then they'd amass on it and start harassing the animal, nipping at its tail and generally pissing it off big time. We all have a breaking point, saltwater crocs included, and at some point it would drop the food in its mouth so that it could snap at the sharks that were harassing it. As soon as it did that, some of the wily sharks would grab the food and scarper.

Later on, we saw this same group of sharks have their own hard work exploited. The sharks nipped and dived around the reef, chasing out prey which would break for the open water, but before the sharks could catch it a pod of dolphins were lying in wait to snap it up. It's a tough old world out there, and some sharks have developed an ingenious way to kill their prey ...

TURNING TAIL

For many years, fishermen were perplexed about why they would catch a particular kind of shark in a peculiar way. Usually when fishing with a bait and line, the fish – or shark – would bite at the bait, and the hook would catch in its mouth. Fishermen couldn't understand, therefore, why one shark species kept coming up with the hook not in its mouth, but in its tail.

The shark in question was the thresher, a beautiful creature with an extremely long tail. Had these sharks forgotten which end of their body was for eating?

Thanks to the work of scientists and underwater cinematographers, it was discovered that threshers have developed an extremely original and effective hunting style, and that this animal's beautiful long tail has a deadly intention – slapping its prey into a stunned stupor.

To do this, a thresher will swim quickly towards its prey. This could be a solitary creature, or a mass of fish swirling together in what's known as a bait ball. Then, in an act of deadly grace, the thresher will flick its long tail over its head, literally slapping the sense out of its prey. The thresher's tail is its own personal stun gun, and once its dinner is dazed and unable to effect an escape, the thresher will finish it off with its jaws.

The thresher is one of my favourite sharks because of this unique hunting style, but also because it's just so adorable. It has a small mouth, big, beautiful eyes, and of course, that graceful tail. I should be careful how much I say about this, however. I don't want to give the *Shark Week* producers ideas. As much as I love the thresher, I really don't want to be volunteered to feel the strength of its whip.

A CHANGING MENU

So what do sharks eat? The simple answer is: lots. This is especially true of the changing diet through a shark's life cycle. For shark pups, they need to eat whatever they can get (while also avoiding becoming a meal themselves). This food could come in the shape of fish eggs, small fish and even other baby sharks. Family dinners take on a bit of a different meaning in the shark world, as even a pup's own mother may see it as prey.

As I mentioned previously, I was once witness to a necropsy on a juvenile great white shark that had been taken from life far too soon because of archaic and barbaric shark nets. When these tragedies happen, researchers try and learn as much about the animal as

This is the sort of image that most people think of when they hear the word 'shark' – a great white with gnashing teeth. But there is so much more to these fascinating and majestic creatures.

I'm one of the unlucky few to have been the victim of a shark attack, in which I lost parts of two limbs – not because that bull shark was a man-eating monster, but more likely because it thought I was something else. But rather than scaring me off going into the water again, it strengthened my feeling that there's really nowhere I'd rather be. And I know that for every person killed by a shark each year, there are 10 million sharks killed by humans. You tell me who the real 'monsters' are.

The great white is only one of hundreds of species of shark in our oceans, but the whale shark (pictured here) is the biggest. They can grow over 60 ft in length and to a scale-busting 30 tonnes in weight. Amazingly, these gentle giants reach this size despite only existing on a diet of tiny plankton that they filter by swimming through the water with their mouths wide open.

At the other end of the scale is the lantern shark, the smallest known species. Living more than a thousand feet below the continental shelf that runs off the coast of Colombia and Venezuela, it has big bulbous eyes to help navigate the continual dark it lives in. As you may have guessed from its name, the lantern shark also has a special trick: it glows in the dark.

Even though there are still hundreds of shark species today, there used to be many more. The frilled shark is known as a living fossil – you only have to take one look at it to know that this animal's near ancestors were around at the time of the earliest dinosaurs.

Another shark that looks too weird to be real is the wobbegong shark, often referred to as a carpet shark because of its near-flat shape, bumpy features and flowing frills that make it appear like the coral in which it lives, hunts and breeds.

(*Top*) Lots of creatures in the animal kingdom have weird and wonderful shapes to attract attention, but this shark's head is all business. Because the head is wide and spread out, it allows the highly tuned sensors within to seek out food – in the same way a search party spreads out in a line to find something.

(*Left*) The epaulette shark has developed one of the most unique and extremely specific hunting techniques out there. When the tide is out and parts of the reef are exposed, it *walks on its fins* between the tide pools to find the 'fish in a barrel' left behind.

A true apex predator is the mako, which is shaped like a missile to hunt the fastest fish and has teeth that angle inwards so it can snatch and hold its prey – evolution has certainly worked in its favour.

The tiger shark is known as the 'trash can of the sea', as it will eat almost anything: birds, octopuses, crabs, smaller sharks, dolphins, camera equipment, rocks, lawnmower engines – you name it. Here, I am showing one that my hand has already been had away by a bull shark. But, to be fair, he'd probably be as happy eating my hook.

Another shark with an extremely original and effective hunting style, the thresher shark uses its long and beautiful tail to slap its prey into a stunned stupor, like a personal stun gun. It's one of my favourite sharks, for its unique hunting technique – and it's just so cute.

One of the most impressive hunters is the spinner shark, an aerial acrobat of world-class standard. They can spin rapidly like a tornado up through a school of fish, chomping as they go, and launch themselves 20 ft in the air – an extremely impressive display to anyone able to witness it (who isn't a fish).

To study sharks up close, we are able to put sharks into a state of tonic immobility, where the mind and body of the shark seem to shut down and the animal remains calm and gentle. This photo shows me doing this to a Caribbean reef shark. It is safe for all involved, and the cameras help us study more of the shark's movements and interactions.

Tagging sharks also allows us to study so much more about their lives – this is a photo of me placing an internal acoustic tracking tag into a big male bull shark (the same species of shark that I fell foul of in 2009). I feel so lucky to get to be a part of this work and learn from scientists who are leaders in the field of shark research.

Getting up close with these amazing creatures is such a privilege. Here I am getting friendly with a blue shark. I'm sorry to say that they are one of the most fished sharks on the planet. I find that incredibly sad, especially as they are so calm and trusting. Some people have had incidents with them while they're feeding, but blue sharks often act like curious puppies, coming straight up to your face for a look.

I wrote this book because I want to share my appreciation of how truly amazing these animals are with anyone who will listen, and especially with those who fear them. I will never stop fighting for them.

they can, and the contents of any creature's stomach reveal a wealth of information not only about that particular animal's lifestyle and diet, but also the species at large.

In the case of this juvenile great white, much of its stomach contents had already been digested, but the more recent meals gave us a solid look into what it had been eating. Some parts of an animal break down in the shark's stomach faster than others, and we found several squid beaks, even though the rest of the animal had already been digested. Squid are dense in protein and a great food source for a young great white, who wouldn't have any trouble in catching them. These juveniles are not picky eaters, and I've even witnessed them taking out birds that were resting on the ocean's surface. I expect part of this is also to help develop the young sharks' skills of attacking, known as predating, from below, as this is what they will need to perfect to get the juicy seals that older great whites tend to base their diets on. I once saw a great white take an entire mutton bird in its mouth, only to then open its jaws and let it fly away – what a story that guy had to take back to his mates!

It's important to remember that when a great white is born it is already 3.5 to 5 ft in length. Juvenile great whites have no trouble taking out fish, and even smaller sharks. As they grow larger they require more calories and nutrients, and this necessitates a change in diet. Fish just won't cut it, and instead great whites start hunting marine mammals, including whales, dolphins, seals and sea lions. The great white is second in ocean predators only to the orca, which will make a meal of the great white to get at its fatty liver. Who would have thought that even the biggest great whites had to worry about becoming someone else's dinner?

THE TRASH CAN OF THE SEA

Every family has someone who will eat whatever gets put in front of them, and for the shark family, that's the tiger shark. In fact, the variety of things that this animal will eat has led to it being given the rather unflattering nickname of the 'trash can of the sea'.

When these sharks are on the younger side they have very pronounced stripes, just like the mammal tiger, and these give them great camouflage on the seabed and reefs. As the sharks grow they have less need to hide, and the stripes gradually fade but always remain visible to a certain degree. Tiger sharks are fearsome predators, and there is little that can harm them in the ocean except for orcas, great whites and, of course, humans.

Tiger sharks will eat almost anything: dugongs (manatees); birds (snatching them from the ocean's surface); arthropods (shrimp and crab); cephalopods (squid and octopus); smaller sharks; dolphins; and turtles (which they are known to sometimes swallow whole).

Some tigers have even developed a taste for cattle. You may be wondering how these animals find themselves in the ocean, and the answer is that humans put them there. Norfolk Island is a tiny island off the coast of Australia, and when the cattle there die of disease or famine they are put out to sea rather than left to rot on land. The tigers in the area soon make use of such a banquet, and they've developed a taste for it. My friends Andy Casagrande and Riley Elliot shot a *Shark Week* show there and went diving with these sharks, and they told me that Norfolk Island has some of the biggest tiger sharks they've ever seen.

So far we've established that tiger sharks have a big appetite for meat, but what does that have to do with being known as a trash can? In fact, that nickname comes from some of the weirder things that they've been known to eat, like trash bags, for one. Yes, tiger sharks will chew through and eat literal trash bags full of junk. They've even been seen eating rocks, and I've witnessed them eating some very high-end camera equipment.

Indeed, tiger sharks have been found with some extremely strange stuff in their stomachs, including a horse's head and a fur coat. (Did they eat the prop department of a mafia movie?) Lawn-mower engines, gumboots, car tyres, a full suit of armour and a porcupine have all been on the menu. In one tiger shark's stomach, believe it or not, they found a whole chicken coop with the chickens still inside!

These sharks do not mess around when it comes to dinner time, snaffling up anything and everything that they come across. Lucky for them they have a highly acidic stomach which can break down a lot, and that's a good thing – imagine trying to poop out a full set of armour.

Of course, in their natural habitat there is very little that a shark would come across that wouldn't be edible (rocks aside). Even for a creature like the tiger shark, eating our trash can be fatal. For now, though, let's take a moment to appreciate the undisputed eating contest champion of the sea, the one and only tiger shark.

NEXT-LEVEL SENSES

Bull sharks are one of the few shark species that can survive in fresh water. Giving birth to their pups in fresh water means that they're kept away from many other predatory sharks, but what do they eat when they're not in the ocean, and how?

Where fresh water meets salt water is usually murky as the different currents mix and churn up sediment. Estuaries and rivers often carry mud with them, and all of this can add to very poor visibility in brackish waters. Bull sharks can sense their prey without the need to see it, and this is enough for them to strike successfully. Can you imagine being able to smell a chocolate bar from over a mile away? How about knowing that a potential romantic partner is in the same part of town without ever laying eyes on them? Sharks can do all that, thanks to their plethora of incredible senses.

Here's where our old friends the ampullae of Lorenzini crop up again. Found at the nose end of a shark, and named after the Italian scientist who discovered them, these are a series of tube-like hollows beneath the skin on the top and the underside of the shark's nose, and sometimes behind its eye. They are similar in appearance to pores on our own skin, but they are filled with a thick, gel-like substance that can detect weak electrical signals. Once a signal is picked up, the ampullae send that message instantly to the shark's brain.

So what kind of electrical signals are we talking about here? As you've probably guessed, the shark isn't trying to pick up Wi-Fi. Believe it or not, all life emits electrical signatures. By having a sense that can detect them, a shark doesn't need to see its prey to zero in on it. If we use a stingray as an example, it may well be hiding beneath a thin layer of sand, but as its muscles move and its heart beats, it will create electrical stimuli. Stingrays are a favourite food of hammerheads, and the wide head of these sharks is loaded with sensors to detect its prey. In fact, the broad head – which is called a cephalofoil – has the largest number of ampullae of Lorenzini of any shark.

It's taken a lot of very clever people to figure out what is going on with sharks and their sensors, and I was lucky enough to work with one of them in the Bahamas, filming a programme called *Sharks Among Us*. Dr Craig O'Connell is one of those intellectuals who can be found anywhere from the rainforest to diving with great whites and is always fascinated with discovering more about animal behaviour. For this show, we were experimenting with supermagnets embedded into fake stingrays, the aim of which was to see if this strong magnetic field would attract or repel the local hammerheads and their highly tuned electrical receptors.

Because a bull shark decided that I was going to be breakfast I no longer have a right hand, and so when I dive I use a carbon-fibre socket that attaches to my forearm, and on the end of this I have a steel hook that allows me to gently fend off sharks who get a bit too close, and to latch onto objects, such as a boat's ladder. I should have foreseen that a supermagnet would attract metal, but then again, I was diving with massive hammerhead sharks and was understandably a wee bit distracted.

Craig and I dived to place the fake stingray down on the ocean floor, and no sooner had we covered it in sand than an 8-ft hammerhead zeroed right in on it. We both began to swim quickly away, but only Craig actually moved – my metal 'hand' was stuck to the supermagnet! Or should I say, I was stuck to what the hammerhead thought was prey.

The big predator had no fear of me, and all I could do was reach out and push its huge head away from me using my left hand. Fortunately for me, hammerheads are very rarely aggressive to humans and the big shark gently turned away to make another approach. I wasn't done worrying, however, as there were bull sharks around, and if they came in thinking I was between them and a meal, then my dive would get a lot more complicated. Fortunately, I was able to finally pull away from the supermagnet, and I'd learned a very valuable lesson: don't mix magnets with metal hands and sharks.

Any tour through the magical realm of shark hunting senses would be incomplete without addressing the shark's nose. You may have noticed that when sharks swim they often sweep their head from side to side. The reason that they do this, aside from the mechanics of swimming, is because it pushes the maximum amounts of water and scent into the nasal passage, which contains sensory cells that enable the shark to smell. Basically, the shark is following an underwater trail of soggy bread-crumbs. With up to two-thirds of a shark's brain dedicated to smell, it's no wonder they're such efficient hunters over both vast and short distances.

As humans we often underestimate just how good our own sense of smell is. When I was in East Timor on peacekeeping operations, we'd spend long periods living in the jungle. You'd be amazed at how in tune we got with our surroundings – you could smell the scent of food cooking, body odour and other man-made scents at distances you never would have believed. This is partly because we were used to being bombarded with so many scents in urban life, but also because in the jungle we took the time to stop and pay attention to what our senses were telling us. A shark's nose works in a similar way to ours, but instead of air flowing into the nostrils, it's water flowing into the shark's nares. This water contains billions of particles of salt, algae, coral, sand, sediment, pollution and fish poop that wash over the olfactory lamellae – folds of skin containing sensory cells – which then send signals to the brain. Odour signals fit into specific receptors inside the nares like a lock and key. This results in the transmission of signals to the shark's brain so it can decipher what signal it is. It's the brain's job to decode these signals, working out things like direction and distance, and whether it's to a mate or food source. Sharks aren't too picky about the latter, however. During a *Shark Week* show called 'Sharkwrecked', where James Glancy and I spent 48 hours in the Atlantic Ocean to display how you might survive after a shipwreck, one of our cinematographers had to answer a call of nature and pooped in the water while swimming and filming. I was a bit worried about where it would float, but a helpful oceanic whitetip swam over and gobbled it right up. Anyway, no one ever said that sharks have great taste in food, and if you've ever smelt decomposing whale meat – a shark delicacy – then you know exactly what I'm talking about.

Although our sense of smell does have some things in common with a shark's, there are some striking differences. Obviously, the fact that they smell via the water is one, and another big one is that their senses of taste and smell are not linked together like ours. If you have the flu and have bunged-up sinuses, it's very difficult for you to taste, although your brain can predict what taste will be like based on past experience. A shark, on the other hand, doesn't know what something it has smelt will taste like until it has it in its mouth, and this is why sharks usually just bite and don't eat humans. They get a taste, realise it's not for them, and go off looking for something more suited to their palate.

TAPETUM LUCIDUM

Sorry for the Harry Potter fans out there, but there are no such things as Hogwarts sharks, and tapetum lucidum is not actually a spell (which is a shame, because maybe if we could turn people into sharks for a couple of weeks, we'd be a lot more understanding of them).

Rather, it is a mirror-like layer on the back of a shark's eye, very much like that of a cat. This 'mirror' increases the intensity of light entering the shark's eye, thus allowing it to see in dark waters. Even in the daylight it can get dark under the sea, and many sharks like to hunt at all hours.

When we filmed 'Sharkwrecked', James Glancy and I were given a cage of sorts to help protect us at night. This was by no means a shark cage, however. It was more like a cheap football net that you put in your back garden, but in a cube shape so that we could float in the middle. It wouldn't stop a determined attack, but it would at least give us some notice that something was right next to us, and we needed all the help we could get – out in the Atlantic at night it was pitch black, and the water was like ink. In such a helpless position, your mind starts to play tricks on you. We'd have sharks around us in the day, but what was lurking there now in the darkness? I took a deep breath and went to the bottom of our 'net' to find out.

No sooner had I flicked on my torch than an oceanic whitetip's face appeared right alongside my own. If I had nares I would have been able to smell its breath, but instead my heart just leapt into my own mouth. It was incredible how this animal knew, despite the dark conditions, exactly where I was going and was there to meet me. I've always thought that diving at night is fun and spooky, but floating in the ocean at night was a little too much.

The next time you pass a mirror, take a look at your eyes. That black spot in the middle of it is the pupil, and it opens and closes (gets bigger and smaller) in response to how much light it's exposed to. So the less light, the wider the pupil needs to expand. Also in our eyes are an iris, a lens, rods and cones.

A shark's eye has these same components, but it is widely accepted now that they do not see in colour. This hasn't always been the case. In fact, it was thought, anecdotally, that sharks paid particular attention to the colour yellow, especially great whites. People even started to referring to yellow wetsuits as 'yum yum yellow!' and not surprisingly, black wetsuits are now a lot more popular (and they look cooler anyway).

For a long time, sharks were thought to be visually poor predators, but we now know this is far from the case. In fact, their eyes are possibly ten times more sensitive to light than our own. With more than 500 species of shark this is of course a generalisation, and there are exceptions to the good eyesight rule. As we discussed earlier, the Greenland shark loses its eyesight to parasites, and whale sharks have denticle-covered eyes and less of a reliance on sight – when you're gliding through the water scooping up plankton, you really don't need great eyesight. For an animal like the thresher, on the other hand, their big eyes give them excellent vision for hunting. At the other extreme, the hammerhead, due to its eyes being widely spaced at either side of its massive head, has nearly 360-degree vision.

It's fair to say that hammerheads may possibly have the greatest eyesight of all sharks. They see in stereo, which means that they see a slightly different view with each eye (due to the position on their wide head). These views then overlap, giving them excellent depth perception. Where they lack great vision is directly in front of their head, which is obviously at the periphery of each eye. I have to be conscious of this when I swim with these animals, and a helping hand to turn their head a little usually does the job if they're swimming directly at me.

Eyes are very soft tissue, with delicate component parts, and they are the weak spot on many animals. A shark is no different, but they have adapted to counter this. Many species of shark will roll their eyes up and back into the eye sockets, and some also have a protective layer that rises from the bottom of the socket to cover the eye when needed, such as during rough and tumble with other sharks, or eating struggling prey. This layer is known as a nictitating membrane and it's found in other top ancient predators, such as crocodiles.

THE DINNER GONG

Have you ever been underwater and tried to shout? It sounds muted, like someone's turned the volume all the way down on your voice. When I dive with my metal hook on my hand, I bang it on the side of my air tank if I need to get someone's attention, but even that metal on metal sound sometimes goes unheard. This could lead us to the natural conclusion that sound doesn't carry well due to the denseness of water, but the opposite is actually the case. The fact that the molecules in water are tighter together than in air allows them to vibrate strongly and carry sound over great distances. This is the reason that whales can communicate over hundreds – sometimes *thousands* – of miles. A sound wave goes out into the ocean, and gradually descends through the water the further it goes. The deeper the water, the colder the water, and cold water slows the sound wave. Eventually a sound wave reaches a depth of ocean beneath the thermocline: it is here that a dramatic change in temperature occurs, and water pressure continues to increase. This part of the water is known as the sound channel, where the sound wave speeds up, and actually begins to climb back to the surface due to the higher pressure. That's a rather convoluted scientific explanation of why, when underwater, noise goes a *long* way. We don't hear the sound waves because it's a different kind of vibration, and not one that vibrates the bone in our human inner ear.

Nature is all about finding an advantage over your competition, your predators and your prey, and so it would be no surprise if we do find out that sharks use sound in some form to hunt or socialise, but right now, we just don't know.

Although there's little debate that sharks don't use sound to communicate, there's no question that they have the ability to hear. After all, they have ears.

Shark ears aren't fleshy things on the exterior of their body like our own. Instead, they have small holes behind their eyes on either side of their head. Like all of a shark's senses, its hearing is exceptional, and this is down to the three inner ear chambers, and an ear stone (small crystals of calcium carbonate) called an otolith. This auditory system doesn't just allow a shark to hear sounds, but it controls its balance and equilibrium – vital components to an animal that moves through the water in a three-dimensional environment to catch its prey, and tries to avoid becoming prey itself.

A shark's hearing and detection are tuned into low-frequency sounds so that it can detect things like the flicking of a fish's tail as it propels itself through the water. Unfortunately, such as when I was finning across Sydney Harbour, low-frequency sounds created by humans can be misinterpreted and a shark will come to investigate the sound. As we mentioned earlier, a shark's senses of scent and taste are not connected. Because of that, the shark will often investigate with its mouth, and this had led to bites on people surfing and splashing in the water. To a shark, these low-frequency vibrations mean fish, and food.

While this can lead to occasional misadventures between humans and sharks, our ability to mimic low-frequency vibrations comes in very handy for a documentary film crew. Sometimes, simply slapping a fin (the diver's word for 'flipper') against the water is enough to attract sharks. After all, fins are developed to mimic the movements of marine animals, so it makes total sense that these would be noises that a shark responds to. During filming in the Bahamas for an episode of *Shark Week* called 'Laws of Jaws 2' we tested whether reef sharks would be attracted to the sound of spear guns being fired beneath the water, or the sound of spears striking the reef. Neither drew much attention. However, when the spear hit a fish, and the fish started to thrash its tail in distress, the reef sharks instantly turned into predator mode, and swarmed onto the injured fish, devouring it instantly.

DEATH METAL

So we know that surfers, swimmers and struggling fish can attract a shark's senses through sound and low-frequency vibrations, but there's a place off the coast of South Australia that brings the sharks in with an entirely unexpected kind of sound ...

Thanks to its colonies of juicy seals, the Neptune Islands of Australia are a well-known hotspot for great whites. I'm sure by now every one of us has seen film or photos of people in shark-diving cages – maybe you've even been in one yourself – but do you know where the idea came from?

The creator of this simple but game-changing innovation is shark legend Rodney Fox. Rodney is a larger-than-life character who has always lived to the fullest. In 1963 he was one of the top spearfishermen in the world, and it was at the Australian Spearfishing Championships of that year that the course of his life would change forever.

During the event, Rodney was attacked by a great white that bit deep into his waist and torso. The power of the bite crushed his ribs, and the great white's teeth did so much damage to the young man that a lot of his insides were now outsides. As though that wasn't enough hell for the day, the shark then bit his arm and started dragging Rodney downwards – a sheer nightmare that I can absolutely relate to myself.

But Rodney would not give up, and he jammed his thumb into that soft spot of most creatures, its eye. The animal released Rodney, who was then rescued. He soon found himself being used as an example of why sharks are dangerous, but he wouldn't have any of it, pointing out that far more people drown in Australia's waters than are attacked by sharks. Like me, he fell in love with the animal that had attacked him and he set about studying sharks: and hence the cage was born. Now in his later years, Rodney has passed the reins of his operation on to his son Andrew, who is one of the world's foremost great white shark behavioural experts.

I've spent many fun-filled days with the Rodney Fox Expeditions crew aboard their boat. We've been in the surface cage together, and all the way down to the sea bottom, surrounded by curious and massive great white beauties. The crew of the boat utilise bait and the blood of fish to bring out the sharks – a process known as chumming – and this has provided thrills for scientists and tourists alike. Thanks to Rodney's operation, thousands of people have had a chance to meet these majestic masters of the sea in person, and have left with the deep sense of awe that comes from seeing these animals in their underwater homes.

Blood and bait is by far the most effective way to draw sharks to a boat and a cage, but the local government only grants two such licences a year. Shark-diving boats are also limited by local government to just three, and one of the Neptune Islands' fleet is run by Matt

Waller of Adventure Bay Charters. Matt's company does not have one of the two blood and bait licences, and so after a bit of improvisation and experimentation they came up with something that finally did the job of attracting big bad sharks.

Playing death metal.

Some would say that the sharks were already in the area due to the blood and baiting of the other two boats, and OK, that's true, but … death metal?

It did work, trust me, and we have the evidence in our *Shark Week* episode 'Bride of Jaws'. The only way we could attract the sharks over to our cage was by placing an underwater speaker just below the water's surface. Then we pumped out the high-intensity, low-frequency sounds of death metal. The great whites loved it and came to investigate, some even trying to take a chomp out of the speaker.

On second thoughts, maybe they weren't fans after all …

THE LATERAL LINE

Our final marvel of the shark senses is one that's unique to aquatic vertebrates. No other group of animals has what is known as the 'lateral line'. This runs along both sides of a shark's body, from its nose all the way to the tip of its tail. It is below the skin and made up of clusters of canals that contain modified hair cells and thick gel, just like the gel that we find in the ampullae of Lorenzini. These hair cells, and the gel within it, are collectively known as neuromasts. They can actually be found all over a shark's body, but by far the densest concentrations follow the animal's lateral line from nose to tail tip.

As to the purpose of the lateral line, think of it as the shark's 'spidery sense'. The lateral line is a multi-purpose array that picks up and detects vibration and movements in the water. Through changes in water pressure and motion, the lateral line tells a shark where prey and predators can be, and it also allows the shark to know the flow of ocean currents, which is important for efficient swimming (it's more effort to go into a current than with it) and migration.

Let's break down the process by which the lateral line works. When there's movement in the water – be that a fish or a current – the displaced water stimulates the shark's hair cells by swaying them. The hair moves within a gel, which in turn stimulates a nerve and sends a message to the shark's brain. The brain decodes the message, giving the shark an understanding of what's going on around it. In real time, the process takes place almost instantaneously.

On top of this, a shark's lateral line also contributes to its sense of smell, allowing it to detect components associated with odour plumes of food sources and mates, from multiple directions at once. Not only is that incredible, but so are scientists. The fact that they've worked out all these fish facts is just mind-blowing. In one experiment, they covered, or inhibited, a shark's lateral line, and even though the shark had full use of its nares (its nose), its olfactory ability (sense of smell) was severely diminished, and its tracking of odour plumes was affected. The startling conclusion? A large part of a shark's sense of smell relies on hairs along its body.

A FISH-EAT-FISH WORLD

We've established that sharks are great at detecting prey to eat. That's great news for them. The not-so-good news is that other, bigger sharks have those same senses, and they're not afraid to eat their own. You think your family dinners can get wild? Well, shark mothers, fathers, brothers and sisters will all eat each other given half a chance.

It's a wild fish-eat-fish world out there on the high seas, and then there's the marine mammals too. We've talked about the orcas who are at the top of the food chain. No one messes with these 'sea pandas', and they will drown, ram, butt and kill other whales, sometimes for hours at a time. With neighbours like that, a shark has to be on top of its game, all the time, to avoid becoming a snack.

Orcas aren't the only thing that will chomp at a shark. Anywhere in the world that you find crocodiles and alligators, you will also find sharks, and these prehistoric behemoths will snatch a shark up as soon as look at it. Many crocs make the rivers and estuaries of coastal regions their home, and the biggest specimens can often be found in these places. These animals are highly territorial, and no one is allowed in without their say-so, even other crocs.

We know by now that there's another animal that loves estuaries and coastal rivers – the bull shark. These animals are very dangerous in their own right, but they're no match for a big croc, and there are plenty of grisly videos of a crocodile enjoying a shark buffet.

Another potential shark predator that is not a shark itself are groupers. The goliath grouper can measure over 8 ft long and weigh over 1,000 lb. Groupers come in a range of sizes and patterns, and some can even change colour, but that cool trick is not much of a consolation for the sharks that they consume. Groupers are opportunistic and fierce predators with toughened lips and mouths. So tough in fact that they can crush the spines of sea urchins – sharp needles that would drive right through human skin and flesh. Groupers are so gnarly that they have been known to eat sharks almost their own size.

Sharks also have to compete with one of the ocean's most intelligent animals, which is also one of the most bizarre to human eyes and minds. It has no bones, it can regrow its limbs, and it can change its shape and colour. On top of that it has blue blood, three hearts, and has been observed using tools. We know these animals as octopuses, a member of the class of creatures called cephalopods. With four pairs of incredibly strong arms and the ability to blend into its surroundings through chameleon-like camouflage, the octopus is a voracious predator and no creature is safe, not even a shark.

There's a great story that encapsulates just how crafty and cunning octopuses are when it comes to hunting. At an aquarium in Seattle, Washington, the keepers kept finding grim discoveries in their dogfish shark tanks – some of the small sharks were missing, while others were killed and horribly mutilated. Was this the work of one of the other dogfish? Was an employee stealing some of the dogfish, and killing others?

To find out, the aquarium keepers set up cameras so that they could film the tanks at night, and sure enough, they found their culprit. As with a lot of large aquarium tanks, several species lived together. Recently, a giant Pacific octopus had been added to the tank, and the keepers were a little worried that one of the sharks would make a meal out of it – after all, sharks love to eat cephalopods. It looked like maybe they were right to be worried, because the octopus took to hiding itself in the rocks, where its ability to change colour and camouflage itself made it almost impossible to see. But there is no hiding movement, and the cameras that the keepers had placed caught something both impressive and brutal.

As the dogfish were swimming past the octopus's lair, a long tentacled arm would strike out with lightning speed, grabbing the shark and flipping it onto its back. As we know, a shark on its back is placed in tonic immobility – a state of near unconsciousness – and with no way to fight back, the dogfish would then be eaten alive. Smaller animals would totally disappear, but the bigger fish were too much of a meal for the octopus, and it would discard these onto the bottom of the tank, where they would be found by the shocked keepers.

If you think that's impressive, some octupuses even escape from their tanks. Sid, an octopus in New Zealand, has made at least five escape attempts, including one where he was later found hiding in a train. Octopuses truly are ingenious and fascinating animals.

Going back to sharks, we've mentioned that they migrate, and these are far from the only animals in the ocean to make such distant voyages. The large, predatory leopard seal makes the journey to New Zealand all the way from their home in Antarctica, and they do so to feast on a new kind of prey. Scientists aren't sure if this surge in leopard seal visitors is actually new, or just newly documented, but either way, they've made a fascinating discovery.

You can tell a lot about an animal's life from its stomach, but obviously cutting it open is very invasive and only takes place on dead animals. The good news is you can still learn a lot of the same by studying what goes into an animal's stomach and what comes out as poop ('scat' is the more proper word). What the leopard seal scat contained was one clue into what these fierce predators were eating. Another clue came from the analysis of visible scars on the leopard seal's faces. As well as that, some of the seals had needle-like spines sticking out of their faces.

It turns out that the spines in the leopard seal's faces were in fact from ghost sharks. These deep-sea cartilaginous fish have a long spine in front of their dorsal fin. It's supposed to deter and protect against predators (evidenced by the fact that they were sticking out of a seal's face), but they don't seem to have been enough to deter the hungry leopard seal's attack (evidenced by the fact that the ghost fish was now seal poop). Technically, the ghost shark isn't a shark, and it is part of a group of animals called chimaera, which are cartilaginous fish like sharks, although it is thought their biological lineage split from their shark ancestor about 400 million years ago.

We've covered a lot of ways that a shark might meet its maker, from the jaws of crocodiles to the beak of an octopus, but all that being said, the chances of a large shark being eaten by a seal or an orca are actually pretty small, and once they have grown to adulthood, the bigger species of shark are on the safe size. Smaller species never get that break, and their entire lives are fraught with danger, where at any moment they could become a meal for something bigger, faster or stronger. It's a tough life in the oceans, and as we are about to see, human beings have made it infinitely harder.

5

ROLL CAMERA, ACTION

Sharks in popular culture

Of all the creatures on our wonderful planet, I would argue that none has captured the human imagination quite like the shark. In fact, sharks have kidnapped our imaginations, held them hostage and led to more nightmares than probably any other animal that's ever existed (a close second would probably be spiders).

Why are we so terrified about being in the water with them? You are thousands and thousands of times more likely to die from falling down the stairs than being eaten by a shark, but do you break into a cold sweat every time you go upstairs? When you're on the stairs, are you in a constant state of fear, worrying what could happen at any second?

For almost all of us the answer is no, but I'm sure most of us have felt that feeling of fear and dread as we step into the ocean. What's out there? Is it hungry? Will it get me?

A new environment can always be a little scary, and so you can understand why someone who grew up in a landlocked city would be wary of the ocean, but I grew up in and around the sea, and I was bloody terrified of sharks. Why?

Humans have a very powerful imagination. Besides, it's not like there's *no* risk of shark attack. I'm living proof of that. It's just that the amount of time we spend worrying about this particular danger, compared with the actual probability of it happening, is massively out of proportion. Here are a few things I did today that could have been potentially dangerous:

- Took a shower (accidental injuries at home account for more than 18,000 American deaths a year).
- Played with my dog (dogs kill 25,000 people a year around the world).
- Crossed the street several times (around 6,000 pedestrians are killed in America every year).
- Worked out at the gym (114 people were killed using free weights or weight machines in America over a 17-year period, during which almost a million more were treated for injuries sustained while at the gym).
- Crossed the street several more times.
- Took another shower.
- Made breakfast (around 5,000 Americans died from choking every year).

As we see from the numbers, any one of those moments had the potential to be fatal, but I didn't lose sleep thinking about them (all right, maybe the thought of 5 a.m. workouts does sometimes keep me awake).

As for deaths from sharks, the number of annual shark-attack fatalities around the world hovers around five to ten deaths most years. In other words, you are 1,000 times more likely to die of choking or crossing the street than you are of being killed by a shark, and 5,000 times more likely to be killed by a dog than by a shark.

When it comes to assessing risk from sharks, our ability to use reason and logic seems to go out of the window. This is often the case with fear, but the truth is a powerful antidote to terror. There wasn't a single time when I went in the ocean that I wasn't worried about being eaten, so what changed? The sharks are just as big and full of teeth as they've always been. It's my attitude that's shifted, which begs the question: what shaped my attitude to begin with? Was I born terrified of sharks, or was that fear put inside of me?

The ocean itself is a daunting and dangerous place. Have you ever heard of a fish out of water? An ape off land is in just such a predicament. Many people have an intrinsic fear of the sea because we know we can't exist in it.

And though you and I now have smartphones, and probably never worry about food, it wasn't that long ago in evolutionary terms that people were prey for animals on a regular basis. Sabre-tooth tigers, crocodiles, even giant birds: our ancestors lived in a world full of danger, and a healthy fear of mother nature is a good thing, because we should respect her and her creatures. It's a humbling and important lesson to know that we are not always number one.

Our sight is incredibly important to us as a species. We use it for almost every part of our lives except sleep. Have you ever just closed your eyes when you're on a busy street pavement? I don't recommend doing it when walking, but that's just the point – we need our sight to see danger coming. Since we became glued to our phones, a lot of people have been killed or hurt because they walked into danger that they would have seen if they'd had their head up.

So what's that got to do with our fear of sharks?

Because in our nightmares we can't see them coming. They come from the deep, and they come fast. Either they take us quickly and we never know it, or worse, we see that fin turn in our direction, cut through the water towards us, and know that there's nothing we can do to stop the attack that's coming.

You get the picture. Not knowing what is below or around is a huge part of what feeds our fear of sharks. If they flew around in the sky over the desert we'd probably be way more chilled about them.

So it's fair to say that we have programmed, instinctive reasons to fear sharks: they're bigger and faster than us, and we suck at moving and seeing in their territory. That's all well and good, but does it account for just how great a hysteria we have surrounding these creatures?

Personally, I think the answer to that is pretty straightforward. In my opinion, no animal, civilisation or human individual has ever been so villainised as the shark, and the reason for that is simple.

Movies.

LET'S REWIND

Before we get into the role that Hollywood has played in shaping the image of sharks in popular culture, it's worth rewinding further to see what place sharks held in the human imagination before they hit the silver screen.

Fishermen used to refer to sharks as sea dogs, which isn't a particularly terrifying name, and suggests that the men were not deathly afraid of the animals. Of course, tales of sea monsters have been around since humans first took to the open sea, but it's quite likely that a lot of these tales involved marine mammals like whales, rather than sharks. Even if a shark is on the surface, there's little to see but a fin. Whales, on the other hand, make an enormous show and noise when they force water out of their blowhole, and they quite often breach the surface and slap it with their massive tails. Whales are colossal animals, dwarfing even great whites. Though sharks are known to breach themselves – witness the aerial seal-snatching antics that made *Shark Week* so popular – this kind of behaviour is really limited to a few spots around the world, and often away from populated areas. Whales, on the other hand, can be seen cruising the coastlines and breaching the surface in a lot of the world's seas, and they probably stirred the fishermen's imagination a lot more than sharks. Ocean swimming for pleasure – and especially surfing – are very modern human acts. It's only when we step into a shark's environment that we can come across them, and so shark attacks were likely totally unheard of to most people until the last hundred years. That doesn't mean that people weren't respectful of them, however. No doubt many a fisherman lost his catch to a shark that snatched it from the line, and in fresh-water deltas like the Ganges, it is possible that there will be some violent interaction between humans and the testy bull sharks who also live in those waters. Perhaps people just had more of a sense of perspective back then. Even a hundred years ago, infant mortality in the Western world was still very high, and healthy people could be struck down by an innocuous cut that happened to get infected. With that kind of real-world worry, perhaps there was simply no time, or will, to think about fanciful causes of death.

Perhaps thinking of weird, gruesome deaths by scary creatures is something that's woven into the human psyche and is inescapable. Every culture on the planet has its tales of creatures in the night. Werewolves, changelings, creatures that are part man, part beast: there is a primordial part of us that is fascinated by things that want to hunt and eat us.

Some stories make it more personal. They go beyond an animal killing us for hunger and make it about revenge. *Moby-Dick* is perhaps the greatest, most well-known example of this in stories that pre-date the big screen but still belong to the industrialised world. In *Moby-Dick*, the white sperm whale does not attack Captain Ahab and his crew so that it can eat them – it does so because it is on a vendetta. It wants to avenge the wrongs that they have inflicted on whale-kind and the natural world at large.

Revenge is a very human trait, and yet it has made its way into many stories about animals hunting man, and perhaps none is better known than the movie that will forever be synonymous with sharks, and the dreadful fear of them.

I am, of course, talking about *Jaws*.

'YOU'RE GONNA NEED A BIGGER BOAT'

When you saw that movie title, did the theme tune start playing in your mind? John Williams's Oscar-winning movie score has become imprinted on our psyche. For the generations of viewers who have watched the Steven Spielberg movie, the sight of a fin moving through the water will forever be synonymous with '*dum dum, dum dum, dum dum*'. As far as a piece of movie music goes, you'd be hard pressed to find something as universally known as the *Jaws* theme. Even the first two beats are enough to conjure up fear, sounding like the heartbeat of a massive, unseen and terrible creature.

A quick internet search for 'most well-known movie theme tunes' finds *Jaws* consistently in the lists of top-10s, which is pretty impressive when you consider that it's almost fifty years old and many tens of thousands of movies have come out over that period of time.

What really stood out about the *Jaws* theme music was that it was so simple, yet so brilliant. In many ways, it is a perfect reflection of a shark itself – streamlined and perfectly honed to achieve its purpose.

For the few people who haven't seen the movie, here's the official logline: 'When a killer shark unleashes chaos on a beach community off Long Island, it's up to a local sheriff, a marine biologist and an old seafarer to hunt the beast down …'

Note the keywords here:

- Killer shark
- Chaos
- Hunt
- Beast

Pretty evocative, right? I guess this wouldn't have drawn as many people to the cinema: 'A swimmer is killed in a rare shark attack. People stay out of the water for a while. The shark migrates on. People go back in the water.'

Jaws is a successful movie because it taps into our primordial fear. In a rom-com you tap into people's desire to love and be loved. In a monster movie – and that's what *Jaws* is – you tap into kill or be killed. *Jaws* triggers our fight-or-flight senses, and as most people don't think they can take on a colossal great white, the overwhelming sense that we get from this movie is fear.

The logline mentions the marine biologist who comes along for the hunt. It's pretty interesting that there's no part in this movie given over to someone defending the shark's right to life. No one is saying, 'It's his home, we shouldn't kill it.' Even the scientist wants to help hunt the shark down so that it can be punished.

It wouldn't be a good movie if it was just nasty old men being killed, and so the shark seems to take a particular pleasure in going after children. Again, on the part of the writer and director, this is a good choice because it taps into our innate desire to want to protect young ones, and our abject terror when we don't think that we can do that. *Jaws* is a masterpiece of cinema because it punches our emotional buttons and makes us feel like we are the ones being hunted, and I think it's important to recognise that as a piece of entertainment, this is art done very, very well.

Unfortunately, art can have unforeseen consequences, and the terror that *Jaws* evoked in many people left the cinema with them and lives on in their minds. Sometimes in life we witness traumatic events that change the way we think forever, and *Jaws* created this kind of trauma through a screen for many people. After seeing this movie, a lot of people could never go to the seaside again without picturing a massive dorsal fin cutting through the water, and angry jaws chomping an innocent child into pieces as his mum watched on in helpless horror.

Although it is hard to pin this on a movie, it is true that after *Jaws* came out in cinemas a lot of shark hunting competitions emerged around the USA. Perhaps this is a coincidence, or perhaps it is at least partly caused by *Jaws* portraying sharks as vengeful creatures. If you think another person – or in this case, an animal – is going to hold a grudge and keep attacking you and your loved ones, the human instinct is to get them first. By making the shark in *Jaws* out to be an animal bent on revenge, it gave the false impression to people that these animals think and act in ways similar to how we do; whereas in reality, a shark just wants to get a meal and move along. If great whites had the ability to process emotions and possessed a desire for revenge, then I'm pretty sure the orcas Port and Starboard, whom we talked about earlier, would be targeted by a great white vigilante posse.

And while we're on the subject of 'revenge', I should stress here that I hold nothing against Steven Spielberg, the film's director, or Peter Benchley, who wrote the book that the movie is based on. In fact, Benchley even said this about his novel: 'Knowing what I know now, I could never write that book today.'

Benchley actually became an advocate for the conservation of sharks. I can only imagine the conflict that he must have felt. As an artist he must have dreamt of touching so many people and holding them enthralled in a story, but on the other hand he saw how some people let fiction become their reality – either keeping them away from our oceans or demonising an animal that can't even conceptualise right or wrong, let alone hold a grudge.

The appetite for sequels was as big as that of the fictional sharks, and three further films followed over the next 12 years. Big, bad sharks were at the core of these sequels but so was the theme of revenge. How do I know that? Because the fourth movie is called *Jaws: The Revenge*.

Jaws 2 has one of the most chilling tag lines in movie history: 'Just when you thought it was safe to go back in the water ...'

In this story, a giant killer great white comes back to the same town and haunts the same police chief who was the focus of the shark's malice in the first movie. I say 'haunts' because these movies almost have a supernatural element. The shark seems to appear out of nowhere, like a ghost, and vanish without a trace. It goes without saying that it tries to eat the sheriff's kid. In *Jaws 3-D*, the police chief's boys are grown up and one of them works at a SeaWorld-esque place that also has a lagoon. One of the boys doesn't like to go in the water but he does it to impress a girl (a very believable part of the storyline). After the initial shark attack they catch a great white, only for it to die in captivity (actually accurate). What's not so accurate is that the great white's mum then turns up on the scene, and she's angry. Not only does she start attacking people in the water, but she uses her shark nose to shatter underwater glass. I don't know how strong that glass would have to be to withstand all of that water pressure, but apparently a great white nose is a powerful battering ram. I won't spoil the rest of the story for you. OK, I will. They blow up the shark.

That brings us to *Jaws: The Revenge*. Unlike the original *Jaws*, this movie got absolutely slammed by critics and audiences. The widow of the police chief – she's now in the Bahamas – finally loses her son to a great white (his luck had to run out sometime). This shark not only has a tremendous thirst for revenge, but apparently a real skill with Google Maps to have followed this family down the east coast of North America. The tag line for this movie was: 'This time, it's personal!'

It felt pretty personal in the other movies, especially when dozens of fishermen were out dynamiting the water to kill the original shark. I think it's fair to say that the franchise ran out of legs, but as an unknown

Hollywood producer once said: 'Give me the same, but different.'

MISSION OF THE SHARK: THE SAGA OF THE USS *INDIANAPOLIS*

We've already discussed the USS *Indianapolis* incident earlier in the book, and it's fair to say that the 1991 movie dramatising this, *Mission of the Shark*, scared the hell out of me when I first saw it as a kid, and stuck with me when I served in the military. This movie doesn't give human qualities to the sharks, like *Jaws* did, but the tension and fear were palpable when the men are in the water and awaiting rescue.

You know what's strange? It's a submarine that torpedoes the ship, killing many sailors and putting the rest at the mercy of the sea, but I didn't ever have nightmares about being sunk by a submarine, just getting eaten by a shark. Submarines killed tens of thousands of people during the Second World War, but it's the stories like the USS *Indianapolis* that really haunt us. When it comes to the idea of death, we seem to have a higher tolerance for stories where humans are the culprits. As a kid I watched so many war movies, but that didn't put me off joining the army. On the other hand, *Jaws* and other films terrified me from going in the water.

The story of the USS *Indianapolis* was remade in 2016; *USS Indianapolis: Men of Courage* stars Nicolas Cage and was made with an estimated budget of $40 million. It grossed little more than $2 million worldwide, so maybe people are tired of shark-attack movies (or Nicolas Cage).

In 2019 *USS Indianapolis: The Final Chapter* was released as a made-for-TV movie. This was not a dramatic retelling of the events, but in fact an expedition to find the final resting place of the ship. The documentary combined archive footage, CGI, expert testimony and survivor accounts, as well as footage of the wreck, which had been undiscovered for more than 70 years. As a veteran, and someone who has benefited my whole life from the freedoms won for me by men like this, I found it a very moving tribute, and my hope is indeed that this story will be remembered for the bravery of the crew, and not the way that some of them died.

We'll talk more about documentaries in the next chapter, but I want to close with one more movie. One that is, perhaps, the most successful shark movie since *Jaws*.

The logline of *Shark Tale* reads: 'Underachiever Oscar (Will Smith) is a pint-sized fish with grand aspirations. When mob-connected great white shark Frankie is accidentally killed, Oscar concocts a story with Frankie's peace-loving brother Lenny (Jack Black) that it was he who murdered the shark. Suddenly hailed "Sharkslayer" by his aquatic brethren, Oscar has bigger fish to fry when Frankie's father, mob boss Don Lino (Robert de Niro), dispatches his henchmen to track down his son's killer.'

Now obviously this is an animated movie, and not to be taken too seriously, but considering that it's a movie aimed at children, I think it's worth looking a bit more closely at what it suggests about sharks. After all, what we see as kids often gives us an impression that we don't shake off in adulthood.

The comparison of sharks to the mafia really speaks for itself. Sharks eat other animals to live, but so do the majority of human beings. Comparing an animal to an organisation that deals in human trafficking, assassinations, torture and extortion seems a little unfair. I get that this movie isn't supposed to be taken at face value, but these little bits of popular culture all add up to form our overall view of sharks, and whether we know it or not, our bias can swing on something that we saw as a kid.

On the flip side, as with any of these films, perhaps this movie will spark kids' fascination with sharks and launch them on a journey of discovery about these incredible animals, and that can only be a good thing. In fact, Will Smith himself was afraid of sharks, and he wanted to dive with them as part of his bucket list. Discovery asked me if I would like to be involved, to which of course I said yes – it's Will Smith!

Taking Will underwater to feed tiger sharks was a great honour, and the man was awed by the experience. Without doubt he came out of the water with a greater love and respect for these animals, and my hope is that movies like *Shark Tale* set many other people, particularly the young, down that same path.

Incidentally, when I was a kid there was a cartoon show called *Sharkie and George*. This was about a pair of fish who went around the sea busting crime, and as you can probably guess, Sharkie was the muscle of the pair. The American cartoon *Shark Squad* took this to the next level, with a bunch of shark/humanoid characters who all looked like they needed to be tested for performance-enhancing drugs. The characterisation of the shark is that of a tough guy, and that's led to its name popping up in a few strange places ...

MORE SHARKS IN POP CULTURE

The Natal Sharks and Sale Sharks are both top-flight professional rugby teams. I don't know if you've ever played rugby, but it's physical and tough, and so I take it as a sign of respect that these teams chose sharks to be their mascots and team emblem. Natal is in South Africa, where great whites explosively breach the sea's surface to snatch seals, but Sale is a part of landlocked Greater Manchester, so I'm not really sure what the connection is there. It does rain a lot in Sale, so maybe the connection is that they're always playing underwater ...

Sharks are a popular subject in books, either as an animal to study in non-fiction, or one to be afraid of in fiction. They make for a great danger to a protagonist when they are either dumped into a shark tank (Ian Fleming's James Bond), or crash into the sea (*The Raft*), and they can even throw an idyllic commune into chaos (*The Beach*). 'Shark-infested waters' are an easy three words for a writer to use to conjure up fear, but let's just stop right there. A shark isn't infesting the water, it lives there. Would you say that humans infest their own houses?

Although there have been tips of the cap to the majesty of these amazing creatures, I think it's fair to say that the majority of movies, and works of fiction in general, have turned the shark into a misunderstood villain, often at times attributing to it the kind of evil qualities that are in fact only inherent in human beings. If sharks had lawyers. they could sue the hell out of studios for defamation of character, but alas, that's not going to happen. All we can do is to try and get the truth out there, and one of our most powerful weapons in that fight is factual documentaries about these spellbinding animals.

6

SHARK ATTACKS

The human stories

While it's true that our impression of shark behaviour has been driven by movies, there are of course instances where sharks have attacked humans, and sometimes fatally. I think it's important that we take a look at them, not to terrify ourselves, but to learn from these incidents. Being attacked by a bull shark was the worst day of my life, but I was able to utilise the lessons from that period to create a whole new and amazing life. Not every attack has a happy ending like that, but there are always lessons that we can learn from them. Instances of shark attacks are already extremely low, but as long as we go into the ocean – their *home* – they will occur, even if it's in low numbers. Nothing in life comes without risk, particularly going into an environment like the ocean.

SYDNEY HARBOUR, 2009

My childhood fear of swimming had not left me when I joined the Australian army, and it certainly hadn't left me since I'd transferred into the navy's elite clearance diving unit. I spent almost every day on or in the waters of Australia, and sharks were never off my mind, but despite visiting their home so often, I'd never come across a big shark. I'd seen a couple of Port Jacksons and wobbegongs, small sharks that were friendly enough, but never the bulls, tigers and great whites that filled my overactive mind. I knew they were out there, there was no doubt about that, but I think we forget at times just how massive the ocean is. Who knows how many times they had come within a hundred metres of me, maybe even closer, but the waters are often dark and murky. It can be hard to see a hand in front of your face, let alone a shark passing through the area. As we've established, they don't rely on sight, and these ocean predators had probably just decided to leave the strange-smelling creature alone.

One morning in February 2009 I was 'finning' across the waters of naval base HMAS *Kuttabul* in Sydney Harbour, meaning that I was swimming on my back and using rubber fins on my feet to propel myself. Like we said earlier, slapping a fin against water creates the kind of low-frequency sound waves that sharks are tuned in to. My fins were going constantly on the murky surface, and that's probably what drew the interest of the bull in the first place. In much of the harbour the water is muddy brown, it was early in the morning and overcast. Light levels were low, and the bull shark wouldn't have been able to see my shape clearly and know that I wasn't a usual food source. Instead, it decided that there was only one way to decide what was making that splashing on the surface:

Bite it.

That was the moment I had dreaded all of my life, and when it happened I was in a state of shock. Yes, my nightmare had been about being eaten by the jaws of death, but to find myself looking into the eye of a primordial predator was just so *weird*.

I was in pain when it grabbed me, but at this moment it wasn't any worse than some of the hard parachute landings I'd had in the army, or skateboard accidents I'd had as a kid. I was hurting, but I could fight. What other choice did I have?

I tried to punch the shark, and that's when I realised my right hand was pinned by its teeth onto my leg. I had to do what I could to get out of this, and so I attempted a counter-attack with my left hand.

And that's when it started to shake me.

All the fight left my body at that moment as the shark's rows of teeth worked through my flesh and bones like saws. We've talked a lot about prey in this book, and I've seen a lot of animals being eaten by sharks when we film, and I sympathise with every one of those guys. If being in a fight with a shark is on your bucket list, take it off. Trust me, you don't want any of that.

The pain was so incredibly intense that it overpowered me, and all the fight went out of me. It was at that moment that I felt sure I was going to die. I remember asking myself if I was ready to leave this world, and when I looked back in those split seconds on all I had achieved from that lost and bullied kid, I decided that I was. It had been a good life, but it was not promising to be a good end, and I started to choke on bloody water as the powerful predator pulled me down into the murk. There was nothing I could do to stop it.

And then it was gone.

I can't tell you why it let me go, but it certainly wasn't because I had fought it off. I was a rag doll in its jaws, but maybe the bull shark had tasted enough to know I wasn't its usual kind of meal. There was a thick coating of blood on the water, and more was pouring out of me. This one animal may have decided I wasn't for it, but how long until more bull sharks came to check out the stink of blood? One bite had almost killed me. Another would surely end things.

Trying to keep my ruined arm out of the water and above my heart to slow the bleeding, I swam for the safety boat and my teammates hauled me in. I saw the look of horror on their faces, so I did what soldiers do, and I cracked a joke. Dark humour in the face of tragedy is often a military fallback to create levity. Then I closed my eyes and prepared to bleed to death.

Doctors, nurses, service personnel and blood donors saved my life. So many people were involved in that day, and my recovery, and it all started with a bull shark. I do wonder what happened to that thing; it may well still be alive. It'll never be able to conceive the role it played in my life.

So what can we learn from that attack? Swimming with fins, alone, across the murky waters of Sydney Harbour is probably not the greatest idea, especially so close to dawn on a cloudy morning. We also didn't have an appropriate medical kit on the boat, and that almost cost me my life – fortunately, one of the lads was willing to shove his hand inside my leg, and held my artery closed with his fingers. Gross, yes, but it saved my life. Many shark-attack deaths come from single bites which result in death by traumatic blood loss. Knowing the position of lifeguard stations, and carrying your own medical kid to surf spots, could be lifesaving. It's always better to have something and not need it, and easy-to-use tourniquets are readily available from camping stores. Of course, a tourniquet is useless if you don't know how to use it, and I think that anyone who is spending time in the ocean should invest in first aid training. Shark attacks are extremely rare, but the ability to perform CPR is a skill that we should all have. The ocean is a beautiful place, but it can be very dangerous and demands our utmost respect. Part of that respect means coming prepared. After all, we are a community, and so taking that time to ready ourselves means that we can be there for other people.

There have of course been a number of shark attacks on humans down the years. One that took place over a century ago directly shaped the course of human attitudes towards these animals, by becoming the inspiration for the story behind *Jaws*.

NEW JERSEY, 1916

The movie *Jaws* was based on the novel by Peter Benchley, but where did Benchley get his inspiration for the story of a killer shark that terrorises a North Atlantic community?

I think it says a lot about just how rare shark attacks are that we need to go back over a hundred years to find a 'serial killer shark'. In 1916, off the coast of New Jersey, USA, one shark is believed to have been responsible for four deaths and one injury. The attacks alarmed the local community and made international headlines. It also made an impression on storytellers, including Peter Benchley.

Visiting the beach on vacation is a relatively new activity for humans, and it didn't become widely popular until the mid-nineteenth century. The summer of 1916 was particularly hot, and without air-conditioning and even fans, people flocked to the beaches to catch a sea breeze, or to dip into the cooling waters of the Atlantic. On top of this, the spread of the infectious disease polio reached epidemic levels that summer, and people believed that their children were less likely to catch it in the fresh air of the coast.

The attacks occurred over a period of 12 days along more than 80 miles of coastline. They began south and headed north. The first victim was bitten on the thigh and bled to death on a hotel manager's desk. There was obvious shock at the attack, but not enough to close the beaches. Fishermen reported seeing several large sharks along the coast, but this wasn't particularly rare: great whites are often found in the waters of northeast America, particularly in the summer.

The second attack occurred 120 metres from shore when a swimmer was bitten around the abdomen. The bite was large enough to sever the man's legs, and there was so much blood in the water that the area around him turned red. Unfortunately for the poor man he was not killed outright, and his screams alerted people on the beach. Lifeguards rowed out to get him, but he did not survive his massive injuries.

The next two attacks took place at a town called Matawan. Thomas Cottrell, a local resident and sea captain, had told people that he'd seen an 8-ft-long shark in one of the town's creeks, but locals dismissed him. This evidence has led many people to guess that the creature must have been a bull shark, but of course, there's nothing to say that the shark seen in the creek was the same one that had killed two people further south.

What is certain is that a group of boys were playing in the creek, and what they thought was a piece of debris floating along the river turned out to be a shark. The boys scrambled to get out of the water but the shark took hold of Lester Stilwell and pulled the 11-year-old under.

What happened next speaks to the bravery of human beings, and in particular the character of Mr Watson Stanley Fisher. After the boys had come to town asking for help, Mr Fisher and several other men dived into the creek to search for young Lester. They believed that he must have suffered a seizure, but no doubt changed their minds when they found his body. In a further shocking turn of events, the shark now returned and attacked Mr Fisher as he was trying to retrieve Lester's body from the water. Lester's body was lost, and Mr Fisher's thigh was so badly injured that he later died in hospital. The young boy's body was finally recovered two days later.

The final attack occurred only a half hour after Mr Fisher was mortally injured. Joseph Dunn, a teenager, was bitten on his leg by the shark, which tried to pull him under. Fortunately, Dunn's brother and a friend caught hold of him, and a savage battle of strength began. Finally, the shark let go of Dunn, who survived his injuries.

The press coverage of the attacks – which had been fairly subdued after the first – went into overdrive once it appeared that one shark was making a habit of targeting humans. Four deaths is of course tragic, but I think we should remember what else was going on in the world in 1916. America had not yet entered the First World War, but for many countries, slaughter of human beings was occurring on a daily basis. The Battle of the Somme, which took place that year, ended with over a million casualties. In New York, a state close to where the shark attacks had occurred, and a city that housed much of the press that ran alarming headlines, there were more than 2,000 polio deaths in 1916. In Ireland that year there were more than 70,000 deaths from tuberculosis, but somehow these deaths do not stir the human imagination as much as the thought of being stalked and hunted and then eaten. Disease kills far more people than animals do, but it doesn't have jaws. Perhaps we need something large enough that we can see to attach our fears to.

Besides the shark attacks, there was another example of this in 1916 America. In Tennessee an elephant named Mary was hanged from a crane after being convicted of murder. Yes, you read that right, an elephant was tried, convicted and hanged for murder.

Mary was found guilty of killing her trainer. As you can imagine, the methods of 'training' circus animals largely involved being cruel to them until they learned to do as they were told to avoid further pain. Mary's trainer had only been hired the day before, and his method of controlling her essentially came down to using a spear of sorts to keep her moving in the right direction. At one point during the circus procession Mary stopped, and the trainer stabbed at her with his instrument. She went into a rage, using her trunk to throw the man, and then stamped on his head, killing him instantly. The crowd was enraged and demanded vengeance.

I won't go into the details of Mary's execution. Suffice to say that what happened to her was awful, and it is evidence that there are few stronger forces in the world than man's urge for revenge.

This same trait of our species was clearly on display earlier that summer following the New Jersey shark attacks. Almost all of the beaches were closed, and people demanded that the government step in. In a year where the world was at war, President Woodrow Wilson somehow found time in his calendar to schedule a meeting with his cabinet to discuss the shark threat, and an offer of $5,000 was made to kill the man-eater, which is a six-figure sum in today's currency. Reminiscent of the scene in *Jaws*, every man with a boat came in search of the man-eating fish. Using hooks, rifles and even dynamite, people took to the waters off New Jersey in search of vengeance, fame and cash.

What followed was a massacre, with hundreds of sharks killed. Many of these were of a species or size that has no record attacking humans, but that didn't matter. They were sharks, and they were hunted down with a fanatical zeal.

Of course, at least one shark was to blame for the attacks, and it had evidently decided that humans were a food source that it should keep going back to. Why this is, no one knows, but one reason could be that there was simply a lot of competition for food that summer, and the thousands of swimmers were the next best thing. We can only speculate about the motive, and indeed, about the shark or sharks responsible. There were two attacks off the coast, and three in and around a creek. Was this the work of one animal? And if so, what species?

A bull shark seemed like a strong candidate. Not only do they swim in fresh and brackish waters like the creek, but they can be extremely aggressive, which could be why one attacked Mr Fisher when he was trying to recover the boy's body (which was, in the shark's mind, its food source).

Witnesses at the attacks had put the shark at between 8 and 9 ft. Given how unreliable eyewitnesses are, especially in moments of high stress, it is definitely possible that the conflicting reports could refer to the same animal.

On 14 July two men caught a great white a few miles from the mouth of Matawan Creek, where several of the attacks had occurred. The animal put up a hell of a fight, and apparently almost sank the boat before it was finished off with a broken oar. Far from being the insanely huge shark that takes down the fishing boat in *Jaws*, this captured animal was under 8 ft long and weighed 325 lb. An inspection of its stomach revealed 15 lb of flesh and bones that scientists identified as human. It appeared that the suspect had been caught.

Some authorities disagreed, mostly on the basis that great whites are not known to frequent fresh and brackish water, such as the creek. Others pointed out that the creek was deep and saline enough that a small great white could well have wandered into it. There is, of course, the possibility that more than one shark was responsible for the attacks. That may be true, but there were no more attacks once the great white with human remains in its stomach had been caught. The official cause of the deaths, listed in the International Shark Attack File, was given as great white attacks.

What makes the events in New Jersey so interesting to me is that they really shine a spotlight on how differently we treat death by shark to death by other causes. Hundreds of thousands of young men were being blown apart by shells and bayoneted by their fellow humans at the time that these shark attacks occurred, but the details of their deaths were confined to a few statistics and headlines. Nobody wanted to know the finer points of how men were killing each other in the trenches, but everyone wanted to know what was killing people in the water, and how they died. The suspected shark – which had actually been caught by a taxidermist – was put on display in a shop window in New York. People wanted to stop and stare at the beast that had caused so much terror.

The other thing that I find so illustrative about this case is the reaction it provoked in our fellow humans. This was at a time when more and more people were taking to the ocean for pleasure, and while there were some people who stood up for marine life, there wasn't much sympathy for the attitude of 'Well, it's their home, and we take chances when we enter it.' There was none of that. An animal had killed a human, and it had to be caught and punished. Of course, that also meant catching and killing hundreds of other sharks who had never even had contact with a human before, but vengeance does not care for such trivial facts as that. 'The Jersey Man-Eater' had given people a stark reminder that we're not always top of the food chain, and as so often is the case when that happens, we lash out in fear, and the results are devastating.

So humans and sharks died, and very little was learned. In fact, the hysterical press coverage would just become the norm when it came to shark attacks – or coverage of sharks in general – and that of course gave rise to the fictional stories that prey on our most base fears.

Fortunately, there are shark-attack stories out there with happier endings ...

HAWAII, 2003

Thirteen-year-old Bethany Hamilton was paddling out to surf with friends in Hawaii when she was attacked by a massive tiger shark. The 14-ft animal severed her arm, but thanks to the quick actions of her friends who got her to shore, and the quick thinking of her father who tied a tourniquet onto her arm, Bethany made it to hospital. Like myself, she was then saved by the skill of medical professionals and anonymous blood donors.

I've spent a bit of time with Bethany, and I take my hat off to her. Within just three weeks of the attack and losing an entire limb, the teenager was back in the water. Just that in itself would be inspirational enough, but Bethany wasn't done.

In 2005, only two years after her attack, Bethany became a national champion in surfing. Not only did she have the courage to get back in the water, but she adapted and overcame the loss of her arm. Anyone who's ever surfed will know about how the upper body is vital for paddling out, duck diving under waves and popping up onto your board to catch a wave. Not only was Bethany able to do this, but she became so good at it that she could beat people who had both arms. What I love most about Bethany's story is that she saw the attack just as an accident and an obstacle that needed to be overcome. Unlike our previous story, there was no desire for revenge. Bethany's experience goes to show that while we must respect the ocean, we should treat it with love, not fear.

The attack on Bethany garnered international attention and even led to movies and documentaries, bringing attention not only to human interaction with sharks, but surfing in general. There's probably no other sport on earth that places humans in and around wild animals like surfing does, and in 2015 that scenario would be played out live on televisions around the world.

SOUTH AFRICA, 2015

'Holy shit! Excuse me.' These were the words of a television commentator as the channel replayed the events that had occurred in the water moments before a surfing event called J-Bay.

J-Bay is an open surf competition that takes place annually in South Africa. During the finals, Mick Fanning was waiting for his turn in the line-up when a great white decided to check him out. The shark thrashed about, seeming to wrestle with Mick's leg rope and Mick gave it a couple of punches to the back of the head in his panic.

To me, this goes to show that very often when a shark approaches a human – either when we're swimming, or on a board – they're just curious. That's little consolation to a victim if that investigation involves being bitten, but I think we need to be realistic here. If that great white really wanted to chew on Mick Fanning, then no amount of punches would have stopped him: the shark would have torn him to pieces. As it was, Fanning's quick thinking and ability to act under pressure did enough to make the shark realise that he was not food, and it swam off and left Mick unharmed.

Like Bethany Hamilton, Mick Fanning didn't let the attack put him off his life in the water, and those who were most terrorised by this attack seemed to be people who were watching it from a very dry sofa thousands of miles from the event.

CAPE ST LUCIA, SOUTH AFRICA, 1942

Earlier we talked about the USS *Indianapolis*, and how that warship was sunk by a Japanese submarine, leading to shark attacks in the water on the surviving crew. That incident was not unique in the Second World War, and in 1942 and 1943 there were two shark attacks that would go down in history as among the biggest ever.

On 28 November 1942 a British troop ship, RMS *Nova Scotia*, was torpedoed by a German submarine while transiting around South Africa. Some 750 souls were lost, and it is estimated that a third of them were killed by oceanic whitetips. Many of them were in fact Axis soldiers and prisoners of war, unwittingly condemned to death by their own side. In what must have been a scene from a nightmare, the German submarine broke the surface to make a battle damage assessment. Its crew were greeted by wreckage, survivors and screams. After taking two men aboard for interrogation, the submarine left the scene. By the time that a rescue ship arrived the next day, all but 190 men had drowned, succumbed to wounds or been eaten.

Almost a year later another troop transport, this time the US-flagged *Cape San Juan*, was torpedoed off the Fijian islands. (I have dived in these waters myself, four years after my own shark attack, and I was amazed at how many sharks I could see.) Many on board the *Cape San Juan* were killed immediately, but it is estimated that 695 died while in the water. It's impossible to know how many of these deaths were down to sharks, but what is for certain is that some of the survivors were still being attacked by sharks as they were pulled from the water by the rescue crews. Without doubt it would have been an absolutely hellish experience.

In contrast to the New Jersey attacks, these incidents were not widely publicised at the time. I suspect that there was so much horror going on in the world that the governments didn't want to scare their sailors and troops any more by having them picturing such a scene at sea. As a serviceman there are always things to be afraid of, but sometimes the less you know the better.

As with the USS *Indianapolis*, both of these ships had been hit by torpedoes, and there would have been an insane amount of blood, noise and movement in the water. It's very possible that the sharks went into a full feeding frenzy as they competed for a meal.

RÉUNION ISLAND, 2015

While Bethany and Mick both survived their attacks, 13-year-old Elio Canestri did not, and the entire story is a tragic one.

Réunion Island is known as being a shark hotspot, with several incidents involving shark attacks, and as such, surfing is banned there on many beaches. Elio and his friends didn't listen to this warning – an impulse I can well understand from my own teenage years – and while in the water he was attacked and killed by a bull shark.

Two years before Canestri's death, in an atmosphere reminiscent of the 1916 New Jersey attacks, residents of Réunion demanded action from their government to reduce the shark attacks; they settled on a cull of 45 bull sharks and 45 tigers. They also banned swimming, bodyboarding and surfing in the sea. Personally, I don't understand the logic in killing animals in the waters that you've just told people to stay out of, and neither did many of Réunion's residents, who were against the culling of these animals.

Prior to Elio's death there had been seven fatal shark attacks off Réunion over a period of five years, and hundreds of residents had protested in favour of culling. Several local politicians, including surfers, now want the island's famous waves to be opened up again. It's a real dilemma. Those in favour of removing the bans say that they have a right to access the water that surrounds their island, but what about the rights of the animals that live in it? I'm a big believer that people should use the ocean, but that comes at your own risk. I think the idea that we should kill ocean creatures to make the ocean safer for us is the epitome of selfishness.

There is one huge salient fact that all shark attacks have in common, and one incredibly easy way to mitigate it. The ocean is the sharks' home: if you don't want to run the incredibly small chance of being attacked by one, don't go in the ocean.

Of course, I don't want you to stay out of the water. Far from it, but if the choice in a particular hotspot is culling sharks, or more surfing, then I'll spare the sharks' lives every time. Sport is important to humans, but we have to remember that such activities are exactly that: *sport*. We can do without them if we have to, but a shark cannot do without its home.

Sam Kellett was a teacher killed in 2014 by a great white while spearfishing off a beach in South Australia, and I have the utmost respect for his parents because, in the wake of their son's death, they requested that the shark responsible not be hunted or killed. I can only guess at the incredible amount of strength and grace it takes a person to overcome that most natural human urge – the desire for revenge. If the parents who lost their son can do it, then it gives me hope that as a species we can start to recognise that there is no malicious intent in these sometimes deadly encounters. Shark attacks are rare, and we need to think of them as accidents rather than murders. With the exception of shipwreck survivors, almost all shark attack victims are in the water because the ocean is a magical place that they love. Sharks are a part of that magic, and we must always remember that we are guests in their home.

7

THE HUNTER IS HUNTED

Humanity's role in driving the shark to extinction

I will warn you now. This chapter will make for uncomfortable reading, and it will sound alarmist at times, but that's with good reason – we *need* to be ringing the alarm, and not just ringing it, but reacting to it, because if we don't, all these amazing animals that we've talked about in this book so far will be gone forever. Bad for them, bad for you and me, and disastrous for future generations.

In this chapter we will talk about how we got to this crisis point, what's being done about it, and what actually needs to be done about it. We will talk about what will happen if we don't take massive action, and what can happen if we do. This doesn't need to be a Doomsday story – there is still time, but it is desperately short. If you think of life in the oceans as a movie, we are right towards the end where the hero needs to step up and act in the face of seemingly impossible odds. If the hero fails – if humanity fails – then the earth will fail with it, and through no fault of its own.

I should say straightaway here that I ate meat and fish until I was well into my thirties, and I enjoyed them. I'm not going to sit here and pretend any differently. Things changed for me when I started to work in the animal kingdom as a documentary presenter. One job took me to Africa, and it was there that I met a special forces legend turned conservationist, Damien Mander.

Damien served in the same clearance diving team that I did and then TAG-E, which is the Tactical Assault Group for Australia's east coast. They were the country's domestic counter-terrorism unit, before going out to Africa to create an anti-poaching team. One day when we were having dinner, I noticed Damien was eating from a different pot than the local rangers. I made a joke about him keeping the best cuts to himself, when Damien said that he was in fact eating a vegan option. I couldn't believe this as Damien's a big, muscular guy, and when I asked him what the reason was behind his decision, he said something along the lines of that he couldn't very well dedicate his life to protecting animals in the wild just to come home and eat them. In other words, he didn't want to be a hypocrite.

That struck a massive chord with me. I'd always hated commanders in the army who tell you not to do something, but then do it themselves, and I didn't want to be like that. It was a bumpy road, but on that day I decided that I was done with meat. Damien didn't bully me into doing it, and I think if he did, that probably would have backfired. And so, I'm not here to convince you, and I'm certainly not going to judge you, but what I am going to do is give you the straight facts about what is going on in our oceans, and how it is driving our sharks to extinction.

WHAT IS HAPPENING IN OUR OCEANS, AND WHY?

During the Ming Dynasty in China, something happened that would be catastrophic to sharks living 600 years later. A new dish was created for the ruling class, based on the belief that if humans consumed certain body parts of strong and ferocious animals, then those traits would be passed on to the eater. The dish I'm talking about is the now infamous shark fin soup.

Some so-called 'healer', who must have been a hell of a salesman, convinced the ruling elite that this soup could cure anything from the common cold to cancer, baldness, bed-wetting and eczema, all while solving a man's erectile dysfunction.

We've all heard about snake oil salesmen. Well, in Asian markets there are tiger penis salesmen, elephant foot salesmen, rhino horn salesmen, pangolin scale salesmen, and far more. Nothing is off the table in this market, and why would it be? It's not like *any* of it works. It is simply remorseless sellers taking advantage of gullible buyers. If they get better – either because they would have done so anyway, or through the placebo effect – they are convinced that the product works. If their ailment continues, they are simply told that they haven't had enough of it and must buy more. This massive money-making scheme has been pushing animals to the brink of extinction and over for hundreds of years, and it's soul destroying. Instead of spending a couple of dollars on Viagra or changing poor dietary habits – which works – someone will spend thousands on a tiger penis – which doesn't. The animal's death serves nothing but lining the pockets of the organised criminals who oversee this global scam at every step. We only have to look at what happened to the buffalo in America, or the rhino in Africa, to know what happens to animals when people are making money hand over fist – they don't stop until they're made to stop.

For hundreds of years sharks were hunted for their 'magical' fins, also a display of stature when served at celebratory dinners like weddings, but the losses were sustainable. It wasn't until the advent of modern fishing technologies and the rise of the lower class to form the new middle class in China, therefore having more money to spend, that shark finning has become a problem on a global scale. Of course, finning is just one threat to sharks among many, and the greed for shark medicine is more than matched by the greed for shark meat in European restaurants. There is no one culture or country to blame. This is a global problem created and maintained by human greed, and one of the greatest threats to our planet is overfishing.

OVERFISHING

The simple definition of overfishing is: taking fish from the sea at a rate too fast for the fished species to replace them. With that in mind, let's start by taking a look at the practice of fishing in our oceans, and how it has changed over the centuries.

Fish has been part of the human diet for at least 40,000 years, but it wasn't until the fifteenth century that deep-water fishing became popular. Up until then, most fishing was done in fresh water, or close to the coast. The main reason behind this is down to refrigeration, or a lack of it at the time. The further someone fished from the coast, the less time they had to get their catch to market. As there was an abundance of fish in coastal waters, it just made no sense to go further out into deeper waters.

In the fifteenth century the use of long drift nets pulled behind ships became popular, and to get around the problem of refrigeration, smaller ships would go between the fishing vessels and the shore, bringing fish to market one way, and ships' supplies the other. This allowed the ships to stay at sea for longer and in deeper waters where there were untapped fisheries teeming with life. As you can imagine, at this time there were no such things as quotas, and the fishermen tried to sell whatever they caught.

Fishing was given another boost with the invention of the steam engine. Nineteenth-century steam-powered ships could drag much bigger nets, and the bigger the net, the bigger the catch.

Then refrigeration truly changed the game for fishing practices. Not only were vessels able to store the catch in a refrigerated hold, but every step of the supply chain up to a consumer's home could refrigerate and keep fish, meaning that what was caught at sea could now make it to a dinner table in even the most landlocked city. This has been devastating to sharks and fish, and I am not exaggerating. For instance, in the last 70 years scientists estimate that we have lost as many as 90 per cent of all large fish species.

Just think about that. Our best-case scenario is that, in less than a hundred years, we have wiped out two-thirds of large-fish populations. And if we've done that much destruction in 70 years, just think where we will be in another 70. Anyone can see that this is not sustainable, even in the most optimistic predictions.

It makes me feel bloody awful to know that for most of my life I have been a willing participant in the problem, and so if you're feeling that way now, please understand that it's OK to make mistakes. We don't know what we don't know until we know, and there has been a deliberate effort to keep this information from us, which we will talk more about later. For now, just trust me, this wasn't your fault, but you can be a part of the solution.

In order to find that solution, we need to take an honest look at the cause of this crisis.

Human beings.

Collectively, as a species we are often our own worst enemy – just look at the wars going on around the world right now for a start – and we are certainly the enemy for virtually every other living thing on this planet. We talked earlier about shark attacks, and how they are portrayed in the media, but for every one person killed by a shark a year, 10 million sharks are killed by humans, which totals over 100 million sharks killed per year.

That's right. Every decade we kill over a billion sharks.

A billion.

Part of the problem is that humans are so good at what we do. Sharks have developed over tens of millions of years to become incredible predators. In far less time, humans have evolved to become the most successful living species on the planet. We live in every place, in every climate. Despite all the wars we start, and starvation and disease, our numbers grow exponentially. We consume, consume, consume. Some of us have the luxury of thinking about the bigger picture, but much of the world lives in poverty. Would you be reading this book if you were worried about where your next meal was coming from? Would you care about the plight of the oceans if you couldn't get clean drinking water for your kids? As human beings we tend to focus on what's right in front of us rather than thinking about the future, and that's just not going to work. It's a bleak realisation that on our current trajectory, the current consumption of fish guarantees that we will destroy all of life within our oceans in the near future. This will also have a devastating effect on the lives of people. Without sharks to eat medium-sized fish, their populations go out of control, which means that the smaller fish they prey on are wiped out. Many small fish eat algae, so without them to control it, algae covers reefs, and when reefs are covered, they die.

Considering that between 50 to 80 per cent of the world's oxygen comes from our oceans, a collapse of shark populations, and its knock-on effects, will be devastating to all life on earth.

There is an extremely simple solution to avoiding this disaster – we stop eating fish – but I'm not foolish enough to think that will happen. As humans, we don't like to give things up. If you look at a country's national debt, we're happy with kicking the can down the street for future generations to pick up. What we do now will impact generations of humans as yet unborn, but apparently we'd rather say, 'Sorry, I'm not giving up fish.'

That doesn't mean that the rest of us shouldn't try and make a difference. In fact, I feel like it places a duty on us to do so. We are all connected on this planet, and we need to help educate others so that they can change their minds and habits, and make up for those who are unwilling to give up fish, such as by protest and campaigning for laws which aim to curb overfishing.

So what does overfishing have to do with sharks? After all, I'm sure a lot of people will say truthfully that they have never bought shark meat in a store or ordered it at a restaurant.

To answer this, we need to understand what bycatch is. There's no nice way to put it, so quite simply, when fishermen use lines or nets to target a specific species of fish, other fish and animals get caught up too. This is known as bycatch, and in an incredible display of waste that is impressive even for humans, these dead animals are then just tossed back into the sea. Sharks are often part of this bycatch.

Large shark species are especially vulnerable to longline fishing (the use of multiple baited hooks on long fishing lines) and make up over half the sharks caught either as bycatch or as fisheries targets. Fishing vessels head out to areas with the highest concentration of fish, and of course this is where large sharks will also go, and for the same reason. Unfortunately, the sharks themselves then become prey.

Bycatch is a very modern problem, stemming from the fact that nets and lines now cover huge areas, often at great depth. Even if countries have laws on fisheries they are often badly managed, and some people just bypass the rules entirely, using illegal nets and ignoring quotas.

The biggest issue with modern fishing is that it is entirely non-selective. If you think about spear fishermen, they go underwater, identify what they're hunting, and then try to catch it. It is a very precise and selective process. On the other hand, when using a net, commercial fishermen are just dragging up whatever happens to be in the area.

There has been pushback against this kind of fishing, which is why a lot of tuna fishing is done with longlines that trail a boat covered in baited hooks. Of course, it's not only tuna who are looking to eat, and so these longlines catch other fish, including sharks, and even dolphins and sea turtles. The fish you are buying may not be endangered yet, but many animals on the verge of extinction may have been caught as bycatch to bring you that meal.

Drag nets are even more devastating. These are dragged by trawlers, and the fine filaments that make up this wall of netting called gillnets are very hard for animals to see. All kinds of fish, marine mammals, cephalopods and even seabirds get tangled up in them. Imagine a giant net just got dragged across your town. That's how this kind of fishing is conducted, and then much of the catch – already dead because of the gillnets – is thrown back into the water. Trawling nets also ruin the ecosystem at the bottom of the sea by dragging through reefs and devastating the aquatic habitat.

Over 300,000 marine mammals like dolphins and whales are killed in fishing nets every year. This is a holocaust that affects *everything* in our oceans.

Bycatch is one of the greatest threats to our shark populations. Out of the more than 100 million sharks that are killed every year, it is estimated that at least 50 million of these animals are killed due to non-regulated fisheries as non-target-species bycatch. In other words, the sharks get killed when people are fishing for other species, and then the dead or dying sharks are thrown back into the sea.

There are bodies set up to monitor and enforce fishing regulations, such as the National Oceanographic and Atmospheric Administration (NOAA) fisheries in America, but even they have documented that shark bycatch in tuna-fishing fleets happens 50 to 100 per cent of the time, and that underreporting of these incidents varies from 9 to 40 per cent. In other words, the situation could be even worse than the disastrous numbers that we already know about.

US fishery regulations are some of the strictest in the world, so if they are having bycatch 50 to 100 per cent of the time, and more if it's not reported, then just think what bycatch must be like by other fishing nations that do not have those regulations. The estimate of 50 million sharks killed as bycatch could be two or three times that already ghastly number.

Then there's international non-cooperation. When Britain proposed new legislation to protect mako sharks, it was blocked by both the USA and the European Union. Can we really have faith in regulatory practices when the bodies that monitor them are advocating for the fishing of threatened species?

The counter argument of these bodies, and that of fishermen, is that shark populations like the mako are harvested responsibly (meaning that they are killed within the number limits that fisheries say are acceptable). What that doesn't take into account is that sharks do not recognise human lines drawn on maps and charts, and they will migrate from the waters of one nation into another, where often there are fewer or even no protections. Many sharks are international animals, and that requires an international solution.

SHARK MEAT: NOT FIT FOR CONSUMPTION

As I stated earlier, there is no one culture or country that is to blame for the eradication of sharks, but some are doing more than others in their contribution to the problem, or the solution. We will look more in depth at this later, but first I think we should study shark meat a little more closely. People very rarely do something unless it has some benefit to themselves, and so I think that perhaps the most convincing argument of why we should stop eating shark is because it's just not good for you.

As a kid in Australia, I grew up eating shark meat. Pretty ironic that while I was having nightmares about being attacked by a shark, I was the one who was in fact playing the role of the predator. The shark meat was called flake, and it was one of my favourite dishes to eat because there were no sharp bones that jabbed into your mouth, or that you had to carefully pick out of your dinner (the reason being that a shark's skeleton is made from cartilage, not bone).

Flake was tasty, easy and cheap. With four hungry kids to feed, my parents were big fans of it. What they didn't know – what no one knew – at the time is that shark meat contains high levels of toxic metals and chemicals. These dangerous substances have accumulated in the shark's flesh from the food that they eat and the environments that they live in. Chemicals from ocean transport, mining, pollution from industry and animal agriculture have all leaked into oceans. Because sharks can't expunge these toxins from their bodies, the toxicity increases over time (and let's remember, sharks can live for many, many years). This process is known as bioaccumulation, and it creates levels of toxins that are dangerous for human consumption. When accumulated over time, they can create severe illness, and even death. I am a big fan of the great motivational speaker Tony Robbins, and he had his own brush with this issue. Tony was big into eating tuna. Like sharks, tuna live long lives and accumulate a lot of metals before they are caught for the table, and Tony's diet of tuna led to him suffering from mercury poisoning. If he hadn't caught the cause of the illness and changed his diet, it could well have killed him.

Let's look at each of these toxins a little more closely.

Lead

I'm sure every one of us has heard of lead poisoning. Some people have even gone as far as to suggest that it played a major part in the fall of the Roman Empire. It's not hard to get this dangerous metal into your body. You can breathe it in, swallow it or absorb lead particles, but no matter how it gets into your body, the health ramifications are the same, and severe. Once it is within our bodies, lead is absorbed and stored in our blood, bone and tissue. This enables an environment of continual exposure to the poison, which can lead to cancer and other maladies. Children are even more susceptible to it, even those babies who are still developing in the womb. Lead has caused children to be born with intellectual and neurological disabilities; for the mother, miscarriage and stillbirth can both be caused by lead poisoning; and it can make both sexes infertile.

Thanks to scientific studies, we know that sharks can have extremely high levels of lead in their meat. In 2013 a study published in the journal *Tropical Conservation Science* found that the larger and older a shark, the more lead was contained in its meat – consistent with the fact that a shark cannot expunge these metals, and so the levels rise with the shark's age and continued exposure to them. In 2021 a study in the Bahamas concluded that *all* the shark meat it had tested exceeded toxic levels for humans. That's right, all of it.

Mercury

Eating seafood is often linked to the most common form of mercury poisoning, which comes from ingesting too much methyl-mercury. Like lead, mercury is naturally occurring and something that our bodies can handle to a certain degree, but since the advent of industrialisation there has been an increase of it in the environment. These toxic mercury levels can build up in a person over time, resulting in learning disabilities in children and motor skill degradation. Adults can develop permanent brain and kidney damage, and expectant mothers can pass the toxic metals onto their unborn children who are, due to their small size, the most at risk from metal poisoning. I am not a doctor, so I can't give medical advice, but what I can say is that if I saw a pregnant mother or a child eating shark meat, then I would want to grab it out of their hand.

Arsenic

I don't know about you, but I've seen a bunch of murder mystery shows on TV where they used arsenic to poison and kill someone, usually over an extended period, by slipping it into the unwitting victim's food. Well, there's no need to add arsenic to shark meat, because it's already there.

In 2014 a study conducted in Australia found extremely dangerous levels of arsenic in shark meat. Many shark species were studied, big and small, but all had arsenic levels well beyond acceptable human consumption standards. You really don't want this stuff anywhere near you, as it can damage the lungs, skin, kidneys and liver, and can lead to heart attack, stroke and cancer.*

This is so well documented that it really begs the question about how shark meat is allowed to be sold, purely from the point of view of humans' best interests. If you need more convincing, a quick internet search for 'bioaccumulation of poisonous chemicals and toxic metals in shark meat' will give you lots of results. How this stuff is on our tables and in our stores I can only guess at, but I imagine it has something to do with money. Sharks are viewed by many as a commodity, and this has led to them being killed for the most weird and disturbing uses.

LOTIONS, POTIONS AND CONCOCTIONS

Remember when we talked about the great white's massive liver? That species of shark is not the only one. In fact, relative to their size, most sharks have large, fatty livers that assist them with buoyancy, as unlike bony fish they lack a swim bladder. The oil extracted from a

* As I said earlier in the chapter, I'm not here to try and convince you into following the same diet I do, but I hope I have given you some reasons to stay clear of shark meat that are for your best interests, as well as the sharks'. Do I want to save the sharks? Absolutely, but I'm pretty fond of humans too, despite our flaws, and the idea of people slowly poisoning themselves is awful.

shark's liver is called 'squalene', and if you've ever noticed how oil rises in water (such as in a salad dressing, or an oil spill at sea), then this process is actually what helps a shark stay buoyant – the oil in its fatty liver literally helps raise the shark to the surface, allowing it to be buoyant without expending too much energy.

Livers are a nutrient-rich organ long prized by hunters, human and animal alike, but in the modern era they are sought after for new reasons.

Industries have attempted to make money out of squalene in all kinds of ways. Most notably, as I write this during the COVID pandemic, squalene has been utilised as an ingredient in vaccines. The reason for this is because squalene is what is referred to as an adjuvant, meaning that it is a substance designed to elicit an effective immune response once the vaccine has been injected into the human body.

Squalene is not a substance particular to sharks' livers, and it can be extracted from various plant sources, such as sugar cane and olives. You may wonder why people would go to the trouble of catching sharks when we can plant olives in groves, but the sad truth of the matter is that commercial fishing is so massive that getting shark squalene is actually cheaper, and as with so much in life, money is the driving force behind the decision making. The fleets of fishing vessels are largely unregulated, even if they are legal, and the demand for squalene means that deep-sea shark species are being harvested to the point of extinction. It seems that then – and only then – the people counting the profits will be forced to consider alternative supplies.

Considering that squalene can be made from plant sources, I really don't see that there is a justification for killing sharks for it, but I'm sure some people would make the case that it's at least being used for vaccines. That may be true, but it certainly can't be said for this next industry, and the part it plays in destroying shark populations in our oceans.

Believe it or not, the beauty and cosmetics industry is complicit in the eradication of our shark species. I bet you've never wondered if you've ever smeared shark liver on your skin or lips, or even used it in your hair, but if you use cosmetic products, then there's a good chance that you have. Squalene is a common ingredient in the industry, and is used in anything from anti-ageing creams, and lotions and gels, to make-up and lipstick. The companies that utilise this shark product claim that the ingredients are natural and organic, which is a nice way to gloss over the death and destruction caused to sharks, unwittingly luring the customer into supporting an industry that is both hideously cruel and ultimately unsustainable. Again, these companies could quite easily switch to plant-based sources of squalene, and some have, but many others choose to put profit over ethical practices. Consumer power is a real thing, and it isn't until we stop buying unethical products in favour of ethical ones that companies will change their behaviour. They do not want to make that change, but they want your money, and by using that we can elicit change.

If you're like me, you probably buy the same products over and over, so it doesn't take much time to go to a company's website or look at a product's label for the information we need about whether or not shark liver has gone into making it. When we choose products that are labelled vegan or plant-based, then we are choosing not to fund the exploitation of sharks and our oceans. If that's not enough to convince you, or someone you know, then think back to what we talked about with the metals and toxins. Much of that toxicity is stored in a shark's liver. Is lead, mercury and arsenic really something you want to be smearing on your lips, or rubbing on your skin? Just like sharks, we bioaccumulate toxins as well. Paints, pesticides, herbicides, fungicides, exhaust fumes, plastic particles, cleaning products, clothing and many other products in our everyday life are bombarding our system with toxins and poisons, but we can make choices to avoid as much as we can, and live a healthier, more in-tune life with our natural environment.

Unfortunately, the use of sharks as a product does not end there. Shark skin and cartilage are in high demand for pet treats and supplements. In some parts of the world, dried-up shark cartilage is sold as a miracle cure for all that ails your joints. It's total snake oil and has no scientific backing at all to support these claims, but unfortunately people hear what they want to hear and buy accordingly. Much like rhino horn – which is made from hair and does *nothing* to cure human ailments – shark cartilage is made up of the same stuff as our ears and nose. Ground-up human ear would essentially be the same powder as ground-up shark cartilage, and just as useless. It's high time we put an end to this ridiculous hocus-pocus stupidity and leave the sharks alone in their homes. When sharks are alive and in the ocean, *that* is where they actually do the most to keep our species healthy and alive, by playing their part in the delicate balance of life on this planet.

HUNTING SHARKS FOR SPORT

Australia is a sports-mad country, and when I grew up I was a very good swimmer, competing all the way through my childhood. Sport is a big part of military life too, particularly team sports, and even though I don't do as much of it in my life now, I am still a big fan of certain sports. Getting together with friends to watch the boxing or the UFC is something I look forward to, and I've been honoured to go on dives with both Ronda Rousey and Mike Tyson.

I know enough former athletes to appreciate what a toll a sport can take on your body, but I think it's fair to say that a lot of them knew the risks going in. You don't play rugby and not expect to break some bones or tear some muscle, and everyone on the field is a consenting adult who wants to be there.

Unfortunately, there are some people who have a very different interpretation of sport, where the other participant is absolutely not there voluntarily.

You've probably seen photos on social media of people posing with dead tigers, elephants and other animals that they hunted 'for sport'. They are hunted for the thrill of the kill (apparently they find killing an animal more exciting than seeing it alive) and often keep a part of it to display, which I guess is their way of making up for some kind of deep insecurity. If you need a set of antlers over your fireplace to show that you're a man, you're probably lacking a lot of self-esteem.

It's very hard for me to write this section without getting angry. While I think that the shark fin industry is awful, I can at least understand that the consumer *thinks* they are getting something to aid their health, even if that's a total lie. And while bycatch is dreadful, I can understand that hungry people will put their family's needs before a shark's. I don't agree with any of it, but I can at least understand the motivation.

Trophy hunting to me shows the very worst side of humans. It is the epitome of ego, and a disregard for anything other than selfish desires. Are we hardwired to be excited by struggling with, and overcoming, an animal bigger than ourselves? You can probably make that argument, but the big difference here is that we don't need to be struggling with them. These animals are not coming into our home and trying to eat out children. We are the home invaders, and this trophy hunting is nothing but selfish, callous and cruel. If the people doing it *actually* wanted to prove how tough they are, then they could just get into a boxing ring with another human being, but they won't, because they are cowards, and shooting a lion, or landing a big shark, makes them feel powerful. In case you couldn't tell, they absolutely disgust me and make me ashamed to be part of the same species.

Unlike the trophy hunting of animals, shark trophy hunting is relatively new. On the east coast of America, the so-called sport fishermen hunt sharks regardless of species, the aim being to catch the biggest animal possible. These morons are now reporting that there are fewer and fewer big sharks to catch, which may have something to do with the fact that they keep killing them before they can reproduce.

In trophy hunting, the bigger the better is the general rule, and as such it removes many of the mature animals from the ocean that are vital for repopulation. Great whites, for instance, must wait several decades before becoming sexually mature, and fewer and fewer of them are making it to this age. In the USA in 2011, more than 2.5 million sharks were hunted for recreational activities. In other words, more than 2.5 million animals were slaughtered to provide a bit of entertainment for people who don't have the guts to put themselves in danger, and who would much rather get their kicks from killing an animal that had no chance of defending itself.

I recently saw someone quote from the Book of Genesis in the Bible, claiming that this was proof that man has dominion over animals – in other words, that they were put on earth for man to use as he sees fit. I find this kind of attitude extremely arrogant. Is it not possible that all forms of life were meant to co-exist?

Ernest Hemingway, who was himself a trophy hunter, said: 'What is moral is what you feel good after, and what is immoral is what you feel bad after.'

I disagree. There is a lot in life that can thrill us that is immoral, and as humans, we have a duty to separate our base, primal instincts from our behaviour. We've probably all been in a situation where it would have felt good in the moment to act violently, but do we do it? No, because we know that feeling good, and being good, are not always one and the same. Think of the slave owner who got a good price for selling other human beings. He would have felt good, but does that make it moral? Of course not. We need to start looking at sports fishing for sharks for what it is – behaviour that falls below what should be acceptable for our species. Thankfully, our standards of what *is* moral, and what is right, have changed over time. I can only hope that humanity comes to its senses with sports fishing and ends this barbarism before it's too late.

SHARKS IN CAPTIVITY

Have you ever seen a shark face to face? I am incredibly fortunate to work in a job where I get to dive with these magnificent animals on a regular basis, but growing up, even though I spent a lot of time in the ocean, I didn't ever come across sharks. As a navy clearance diver I saw a couple of the smaller kinds, but it wasn't until a bull shark got me in its mouth that I really got acquainted with one in the wild. Before that, like almost every other human, the closest I'd ever got to seeing these animals was in an aquarium.

Different people will have different opinions on this, but it's useful to look at the history behind keeping sharks in captivity. The first thing to say is that, despite the incredible work of scientists over the decades, there is still so much we do not know about sharks. This isn't because these experts have been slacking in their work, but because sharks are just so complex, and much of their lives so mysterious, that there is so much to learn.

One of these mysteries is about why sharks do not fare well in captivity. Attempts have been made to put many of the more than 500 shark species into captivity, but for a long time only the benthic species of sharks seemed capable of making the transition – a group that includes catsharks and leopard sharks. As our knowledge of sharks increased, this has allowed larger species of sharks to be successfully kept in aquariums, including such species as the grey nurse and whale shark.

Many aquariums are situated hundreds of miles from the sea, and so the transport of the animals is often the most dangerous time for them. A lot of these sharks have been caught accidentally, and some fishermen want to get the sharks to experts who can perhaps save them, while others see a chance for profit, like the Turkish fishermen who put two great whites in a tiny restaurant tank. Not surprisingly, both of the animals died in little over a day in what must have been a cruel and lingering end.

In fact, no matter the conditions, great whites just do not take to captivity. Even when they are kept in large holding pens in the ocean, the animals refuse to eat. Like a lot of us, I did not cope well with the experience of being locked down during COVID. I have always been someone who needs to be on the move, as though it's part of my DNA. I can only imagine what it must feel like for a migratory animal to be kept from fulfilling its natural behaviour.

The closest that anyone has come to housing a great white is at the Monterey Bay Aquarium in California, where they held several juvenile great whites over a period of more than a decade. The longest that an individual shark was kept for was 198 days, but the keepers began to worry about its health and released it back into the ocean.

Predators like bull sharks can sometimes be found in the biggest aquariums, and grey nurse and lemon sharks are also common. After my shark attack in Sydney Harbour, I became quite a news story. It was the first attack there in 60 years, and I was a military service person to boot. I gave interviews to all kinds of programmes and papers. When I decided it was time to get back in the water, and I tried surfing on the popular Bondi Beach, I was recognised by the paparazzi and the photos were used by the media. I've told the story of my recovery in my book, *Uncaged*, so I'll just say it caught the attention of a lot of people, including the producers of *60 Minutes*, which is one of the biggest shows in Australia. Not only did they want to interview me, but they asked if I wanted to swim with sharks in the Sydney Aquarium.

I said yes, but I won't pretend that I wasn't scared. In fact, being scared was one of the reasons I knew I had to do it. I'd spent so much of my life in the water that I didn't want to give it up, and anyway, with the chances of a shark attack being so low, what was the risk it would happen to me twice?

Now, on the subject of animals in captivity, I'm not a fan. I generally can't stand zoos or aquariums. To me, they're just animal jails making a profit out of misery. However, I think that there are exceptions to the rule in certain circumstances.

If an animal isn't safe in its natural environment, and a habitat is created to provide safety while also giving it adequate space to have a sense of freedom, then I can see the benefit to captivity as opposed to letting that species go extinct. This is a human solution to a clearly human problem. It's a very sad reality in our world that we are the driving cause of habitat loss and animal extinction on this planet. Outside of this last-hope scenario, I would agree with short-term captivity for injured wildlife while receiving medical treatment and recuperating, as long as they are reintroduced to their home environment afterwards; or long-term captivity if release isn't an option due to injuries.

I have not, in my experience, seen sharks in captivity fitting into these scenarios. That's not to say that one doesn't exist. I believe that, for the most part, aquariums keep sharks in tanks under the guise of public education and entertainment, but the true reason is to make profit. While there is an argument that these establishments help research, I do not believe that this is a valid reason when that form of research could be carried out at sea.

At the time I had not yet arrived at this way of thinking, and so in I went into the Sydney Aquarium with the grey nurse sharks, and though I will always believe that animals should be free from captivity, it was a very important moment for me. Those sharks helped me skip years of therapy, as they showed me that not only were they not animals to be afraid of, but that they were something to be in awe of. Seeing them move so gracefully and inquisitively through the water, I just knew there and then that my future wasn't just about being back in the water – it was about being back in the water with sharks. That dream became a reality, but the more I got to know about sharks and our oceans, the more I realised that my own injuries were the least of our problems.

8

SAVE OUR SHARKS

We all have a part to play

When it comes to saving our shark species – and indeed, saving our oceans – there is no one solution. Rather, it will involve a series of changes at government and individual level, and huge shifts in culture. It is a massive task, but to fail in this mission does not bear thinking about. When I was going through the rigorous selection to become a clearance diver, my brain kept coming up with all kinds of reasons why it would be a good idea to quit, and so to protect against that voice in my head I had gone into the training with the attitude that I would never quit no matter what. I would either pass the course, or die trying.

Now I'm not saying that we need to give our lives in order to save our oceans, but we may need to give a part of our lives. I know a lot of people reading this probably love eating fish – and I certainly did – but if giving something up for a greater good was easy, then it wouldn't be a sacrifice, would it? Some of the changes we need to make are hard, but we really don't have a choice. Even if someone wants to put their own appetite or profit before sharks and ocean life, then what about their children? Do we really want to condemn them to living on a planet with barren seas? Because make no mistake, that's exactly where we're heading.

So far we've talked about what sharks are, why they're so amazing, and now what threatens them, but I don't ever want to talk about a problem without talking about solutions, and that's what this chapter is about: solutions. I am critical, but I am an optimist. Humanity has made huge strides in changing some of its worst behaviour, and that gives me reason to hope that we can do the same with our oceans, but the time to act is short, and the cost of failure is almost beyond our comprehension.

I also want to reflect on what we can do as individuals, because controlling what we can control is something we can action *today*. It's instructive to look at several countries in the world to see how they are attempting to solve this problem – if they are even trying at all. The reason it's important for us to understand this is because as individuals we have power, but when we come together our power is greater still. If we know what our governments are doing, or not doing, to protect sharks and our oceans, then we can apply support, or pressure, accordingly. If we are lucky enough to live in a time and place with freedoms of speech and protest, then let us use those to protect animals who cannot speak for themselves.

UNITED STATES OF AMERICA

The National Oceanic and Atmospheric Administration (NOAA) is responsible for the management of US fisheries, and therefore the protection of sharks in their waters. During research for this book, I went to NOAA's website to look at their overview of the Shark Conservation Act of 2010, and their opening paragraph is quite telling:

> Sharks are among the ocean's top predators and are vital to the natural balance of marine ecosystems. They are also a valuable recreational species and food source.*

They're spot on with that first sentence, but the second? A valuable recreational species? What kind of Orwellian doublethink is going on at NOAA? They say themselves that sharks are vital to the natural balance of marine ecosystems, then in the next sentence justify fishing for sport?

Unfortunately, this is just one example of the hypocrisy which is so common in these government organisations. If you really wanted to protect sharks, then you'd think banning the hunting of them for sport would be the first item on your agenda.

That isn't to say that NOAA, and the US government, haven't put any measures in place. The Shark Finning

* https://www.fisheries.noaa.gov/national/laws-and-policies/shark-conservation-act

Prohibition Act of 2000 added shark finning to the already existing Magnuson–Stevens Fishery Conservation and Management Act (MSA). The shark finning act makes it illegal for anyone in US jurisdiction – that is, in US territorial water or ports – to possess shark fins without the entire shark carcass. The reason that this is important is because many shark fin fishermen simply slice a shark's dorsal fin off, and then leave the dead or dying animal in the water. They do so to save space on the ship, and the idea of this act is that it makes the harvest of fins far less wasteful. Of course, what it does not tackle is the fact that even if every part of the animal is used, it is still one less shark in the ocean, and that is no good when shark populations are declining because of overfishing, with sharks unable to repopulate at the rate we kill them. The shark finning act is like telling a bank robber that it's OK for him to take gold from the vault so long as he takes all of it. And even when there is a will to police the act, it's extremely difficult. The ocean is a vast place and enforcement resources are limited.

In 2010 the US Shark Conservation Act introduced the requirement that all sharks be brought to shore with their fins attached. Again, while this cuts down on waste, it does nothing to tackle the problem of overfishing, sports fishing and bycatch.

NOAA's website says that 'by conducting research, assessing stocks, working with US fishermen and implementing restrictions on shark harvests, we have made significant progress towards ending overfishing and rebuilding overfished stocks for long-term sustainability.'* However, there is no data provided on the page to back up these bold claims – claims which, by the way, I would love to believe.

The United States is of course the foremost economic, cultural and military power in the world, and as such it can influence fishing far beyond its own waters. As listed on their website, some of NOAA's international work includes promoting 'our fins naturally attached policy overseas and [we] provide technical support for other countries' shark conservation efforts. Support activities include shark identification training and data collection.'

Again, the emphasis seems to be on stopping finning, rather than not killing sharks, and if I had to take a guess at why this is, it's because Western countries are quite happy to ban something they don't really benefit from, but they're unwilling to ban the catching of sharks in general, because that brings in money and satisfies some people's tastebuds. It's quite appalling hypocrisy. What NOAA is essentially saying is: 'Killing sharks is fine so long as it's for a reason we agree with.'

The idea that we can manage our way out of this, short of a moratorium on and cessation of all fishing to allow stocks to recover, is just delusional. I understand, however, that NOAA is a government organisation, and they answer to the federal government, and the representatives at the federal government answer to a lot of people, including their constituents, but more importantly lobbyists. These are the people who are paid by

* https://www.fisheries.noaa.gov/international-affairs/shark-conservation

companies to go in and influence politicians into voting a certain way, and the figures involved are massive. Food and beverage lobbyists spent over $27 million on influencing politicians in 2021, with Pacific Seafood Group spending $230,000, and Trident Seafoods $96,000.*

Money talks, and it drives a lot of policy making, but that's not to say that no progress has been made in the fight to protect our sharks. America is bordered by two massive oceans – the Pacific to the west, and the Atlantic to the east. The Highly Migratory Species Fisheries Management Plan (HMS FMP) covers the Atlantic and bans outright the landing of 19 shark species – which is much more in tune with what we should be doing.

We mentioned CITES earlier, and how some species of shark have been included in their Appendix II, which lists animals not necessarily threatened with extinction at the present moment but could easily be if people were allowed to make a trade out of the animals. Some examples of sharks on CITES include the porbeagle, whale shark, basking shark, hammerhead shark and oceanic whitetip shark. CITES also has an Appendix I, which prohibits *any* commercial trade in a species. In my view, we need to start getting more shark species onto this list, and ASAP. I have included CITES in the US section of the chapter because if anyone has the power and influence to make this happen, it is America. All that is left to see is if they have the will, and there are encouraging signs that this is the case.

Different states in America have also issued their own protections for certain species. The great white, for

* opensecrets.org

instance, has been protected in Californian waters since 1994. Thirty of America's fifty states have a coastline, and the more that follow California's lead, the less the solution needs to be found at a federal level. Of course, with many shark species being migratory this is not a problem that can be totally solved at the state level, or even the federal level, but what starts in one state can often gather momentum and spread to other parts of the country. Changing attitudes in America, particularly towards sports fishing for sharks and eating contaminated shark meat, could shift the needle significantly in the direction of shark conservation. America has been an international force for good in its history, and I really hope that we can look back on its intervention on the decline of shark stocks and say the same thing.

UNITED KINGDOM

'Rule Britannia, Britannia rule the waves!' It's fair to say that Britain's history is tied to the sea. From Viking raids, to conquering fleets, beyond the battle of Trafalgar and the trade of the East India Company, the United Kingdom would not be what it is without our oceans. What better way to repay that gift than by becoming one of the leading countries to save them?

Let's start with some good news. In British waters, it is illegal to kill, injure or possess basking sharks and angel sharks. There are also more restrictions in place to protect these species, but those are the most important. The UK has also become the first country in the world to ban the import and export of shark fins, or products containing them, which is obviously great news. However, we must not be naive and think that just because it is illegal to supply something the problem will go away. As we have seen with ivory and poaching around the world, criminal organisations will go to great lengths to supply customers when large amounts of money are involved. Park rangers in Africa have lost their lives in defence of the animals they protect. A ban is only as good as its enforcement, and so we must face the very real possibility that making shark fins illegal could come at a cost of human lives as these laws are upheld. I wish it could be different, but it's a realistic part of the equation that needs to be considered, so that we know what we are accepting. Thousands of people are killed every year in the war on drugs. It would be incredibly naive of us to think that a global ban on shark fins would not also create an underground market which could itself be very violent. That is why ending the demand is vital, as there will always be people willing to supply it as long as it remains.

Although I am obviously a supporter of this shark fin ban, and grateful to the people who have made it happen, as in the United States it still does not stop sharks being killed. As the UK's animal welfare minister, Lord Goldstein, said: 'Shark finning is indescribably cruel and causes thousands of sharks to die terrible deaths. It is also unforgivably wasteful.'*

* https://www.gov.uk/government/news/government-to-introduce-world-leading-ban-on-shark-fin-trade

Again, the emphasis of this ban is that finning is wasteful, rather than killing sharks is unsustainable. As this is a new law, it remains to be seen how stringently it will be enforced. The Royal Navy and other navies from around the world combined efforts to curb piracy in the Gulf of Aden and the Indian Ocean when commercial shipping and the profits of those companies were at stake – will they do the same to protect the sharks, as they have set out in law? Without a credible deterrent it is hard to see illegal fishing stopping. There's just too much money at stake, but every step forward is a step to be applauded, and I have no doubt that the ban on importing tins of shark fin soup into the UK will make a difference. Ease of access is definitely a consideration here, and if someone can't pick up shark fin from their local shop, maybe that will break the habit of them buying it and introducing it to younger generations in their families. Shark fin soup is very much a cultural food, and if we can change the culture, we can change the demand.

So kudos for the work that the British government is doing on this issue, but there is still a long way to go. For instance, sports fishing is not just something that occurs in America, and on a quick internet search I found company after company in the UK who were offering deep-water shark-fishing expeditions. Some of them are even more specific about what they're going after.

With an online search for shark fishing in Cornwall you can see brave men posing with the monsters they have just struggled with in the deep. That's not the reality. What you really see is a bunch of smiling idiots who have just pulled a majestic creature out of its home and ecosystem at zero risk to themselves. The defence of these companies – and many in America – is that they operate a catch and release system, meaning that they put the animal back into the water after they've dragged it around with a hook in its mouth and posed for a photo.

I'm aware that there is a certain amount of hypocrisy in me saying this, because I have been a part in landing sharks. In fact, part of the reason we know that some animals die after catch and release is because it has happened during shark research. That is why scientists are continuously working on ways to study the animals in the water, so that they never have to leave it. It's not always possible, but that's the aim, and the decision to take an animal out of the sea is never taken lightly – and it's definitely not done just to get a thrill out of landing a shark. You may be saying, 'Why can't we just leave them alone entirely?' and I definitely see your point; I've asked that myself, but what we learn from researching sharks allows us to better understand how to protect them. For instance, by placing tags on animals, we learn their migration routes, and those routes can then be prioritised for protection.

While laws in Britain are moving in the right direction, albeit too slowly to save millions of sharks, there are disturbing signs that attitudes towards hunting sharks are not changing. One magazine article that I found claimed that the popularity of fishing for sharks was growing year on year.* They didn't support

* https://www.rokmax.com/blog/guide-to-uk-shark-fishing-in-2021.aspx

the claim with numbers, but the number of shark fishing expedition businesses would seem to support that idea.

But look, let's be optimistic here. It wasn't long ago that fox hunting with hounds was legal in the UK, but due to a public outcry at the barbarity of the 'sport', it was outlawed in 2005. Theresa May, the former prime minister, had come out in support of fox hunting, but the Tories never pushed ahead with trying to repeal the law, and that's probably because they knew it would cause an outcry from the public. An opinion poll carried out in 2016 found that 84 per cent of people in England and Wales wanted the ban to stay.* I think that is incredibly encouraging. In this day and age, it can be hard to get two people to agree that the sky is blue, but more than four in five people agreed that fox hunting is wrong and needs to remain illegal. Now sharks do not have the cute, fluffy faces that foxes do, and they have a public image problem, but I really think this is a cause for optimism. The British public have shown that they do not support barbarous blood sports, so with the right amount of exposure and activism, I am fully confident that we can get this support behind the banning of shark fishing, too.

The counter argument to the ban is that fox hunting creates jobs. But every immoral trade in history created jobs, and is that really a reason to keep them? The slave trade created work for shipbuilders. War creates work

* https://www.discoverwildlife.com/animal-facts/what-is-fox-hunting-and-why-was-it-banned/

in armaments factories. Society, and our planet, might well be better off without some occupations, and besides, we need people to protect our oceans. What would be a more beautiful story than having the people who hunted sharks becoming the ones who protect them?

It's happened before with sharks. In the Philippines, hundreds of thousands of tourists flock to see whale sharks. Filipino fishermen, who used to make $2 a day from fishing, now make more than $60 a day acting as the whale sharks' guardians.* Not only do they protect these animals from finning, but the fishermen take tourists onto and into the water so that they can see these graceful creatures in their home. Could a similar venture exist with basking sharks, blues and porbeagles? Can we find a way to enjoy the majesty of these animals that does not involve dragging them on a line?

There are more than 70 shark species recorded in UK waters, and a quarter of them are listed as vulnerable on the IUCN Red List, including five that are critically endangered. Great Britain has taken important steps to protect them, but there is still a very long road ahead, and the time to act is short. If you live in Great Britain and you would like to help, then the Shark Trust organisation is a great place to start.

* https://www.scu.edu.au/engage/news/latest-news/2019/poor-filipino-fishermen-are-making-millions-protecting-whale-sharks.php

CHINA

As we know, the depopulation of sharks is not down to any one culture or country. Every nation shares part of the blame, and every nation can be a part of the solution. Shark finning is regularly brought up in legislation, but as we've discussed, bycatch and the deliberate targeting of sharks for meat are always huge issues and are responsible for the decline and endangerment of shark species.

That being said, China – and more specifically, the trade in shark fins – does account for a massive number of sharks killed. As China becomes a stronger economic power, individual citizens have acquired more spending power, and this is partly responsible for the increase in shark finning.* It is a billion-dollar industry, and as we stated earlier, supply only occurs when there is demand, and demand in China has been sky high.

A study in 2016 reported an 80 per cent drop in shark fin sales in Guangzhou, China, which is obviously great news. Of the 1,600 residents interviewed, 85 per cent said they had given up shark fin soup in the previous two years, and 62 per cent said they had done so to protect sharks.† I think this is absolutely brilliant, as it shows a big change in culture not because people were told not to do something, but because they think it's the *right* thing to do.

* https://www.ncbi.nlm.nih.gov/pmc/articles/PMC5042495/

† https://www.ncbi.nlm.nih.gov/pmc/articles/PMC5042495/

With estimates of shark finning killing between 26 and 75 million sharks a year, any decrease in its popularity means big numbers of animals saved in the wild. In Beijing, 19 out of 20 restaurants reported that they've seen a 'significant decline in shark fin consumption' in the last three years.* That is huge, and a big cause for optimism.

In China there have been public awareness campaigns and partial bans on shark fin soup, and this has led to more discussion and changing attitudes on the consumption of sharks. The toxicity of the meat is also turning people off from eating these animals.

In 2020 China made the first major revisions to its fishing regulations in 17 years. The aim of this was to curb illegal activity, increase transparency and improve sustainability in commercial fishing. With a fishing fleet of close to, if not more than, 3,000 vessels, China's annual catch dwarfs that of the rest of the world, and as such this one country has a massive effect on fish and shark stocks, making up around 15 per cent of all the world's catch.†

Illegal fishing is a different story again. According to a Pew Trust report in 2018, one in five fish is caught illegally. As you can imagine, those breaking one law at sea are happy to break others, and illegal fishing is responsible for a massive amount of bycatch and the landing of protected species.

* https://www.speakingchange.org/files/SharkReport_Evidence_of_Declines_in_Shark_Fin_Demand_China.pdf

† http://www.fao.org/3/ca9229en/online/ca9229en.html#chapter-1_1

And it's not only the fish that need protecting. Very recently, a Chinese tuna fishing boat was also finning illegally. During that time several of the crew got sick with an unknown illness, and some even died. Three of these men were dumped over the side of the boat. In this illegal industry life is cheap – both animal and human.

China is currently ranked worst in the world for illegal, unreported and unregulated fishing (according to the IUU), but I'll close this section with some encouraging statistics: between 2011 and 2014 the price of shark fin fell by 50 per cent, a reflection of the far lower demand. And a final cause for optimism: in 2016, a study by WildAid of Chinese residents found that 93 per cent had not consumed shark fin in the previous six years. On top of that, since 2011, there has been an estimated 50 to 70 per cent decrease in shark consumption in China. With a strong government, economy and a navy to police its policies at sea, there is no reason why China cannot quickly turn things around, and tens of millions of sharks can be spared this senseless slaughter.

AUSTRALIA

I was born and raised in Australia. I have lived in many parts of the country, I have served alongside my brothers and sisters of the Australian Defence Force, and I wouldn't want to be from anywhere else. And so, what I say now is from a place of love.

When it comes to sharks, Australia is getting it all wrong. In fact, some of our behaviour towards them is nothing short of a national disgrace.

So why is keeping sharks in Aussie waters alive important? Australia has some of the most amazing coral reefs in the world, and these are vital for producing oxygen. Nature and life have developed a finely tuned balance over hundreds of millions of years, and our methods of fishing are akin to hitting this finely tuned machine with a sledgehammer.

Let's start with the good. Australia's waters could, and should, be some of the most thriving shark habitats on the planet. There are all kinds of shark species off Australia's coast, including great whites, zebra sharks, tiger sharks, grey nurse sharks, whale sharks, bull sharks, great hammerheads, Port Jacksons, thresher sharks, pygmy sharks, and the incredible and unique wobbegong, which looks like a cross between a fish and a really tacky carpet.

No country on earth has a higher diversity of sharks than Australia, but one in eight of our species are endangered. Some researchers have estimated that to protect these threatened species would cost the Australian taxpayer about $114 million a year, which at their time of writing was 0.3 per cent of Australia's military defence budget.* Having served in both the Australian army and navy, I am a staunch believer in giving our servicemen

* https://theconversation.com/the-real-reason-to-worry-about-sharks-in-australian-waters-this-summer-1-in-8-are-endangered-161352

and women the equipment they need, and a good standard of living, but we need to ask ourselves the question: what is the defence force for? Personally, I believe that protecting Australia extends to protecting Australian oceans, including its inhabitants. Perhaps this comes directly from using the navy, or perhaps it comes from moving money away from the defence budget and into conservation. I'm not someone who thinks that the world is all sunshine and rainbows, and I believe the need for a defence force exists, but there is more than one way to protect your homeland.

Let's take a look at the current state of shark conservation legislation in Australia before discussing what needs to change. Like many Western countries, Australia has a ban on live shark finning, which is the practice of cutting the fins from an animal while it is still alive, then dumping it back into the sea to suffer a lingering death. What is not banned is the selling of fins altogether. In Australia it is legal to land a shark, then sell its meat and fins separately, with the cartilaginous fins selling for a much higher price than the meat (with its toxic metals). Loopholes in the law are found and exploited by those who want to cash in on this lucrative trade.

Like the USA, different states in Australia have added their own laws for shark protection. For example, great whites are a protected species in South Australia, with an act that prohibits them being taken, harmed, harassed, sold, purchased or possessed.* The penalty for breaking

* https://www.pir.sa.gov.au/fishing/sharks/fishing_restrictions_for_sharks

this law is up to a $100,000 fine, or two years in prison. Given how much money can be made from a kilo of shark fin, you can bet that some fishermen – particularly those coming from outside of Australia – are willing to take that risk. After all, Australia has harsh punishments for the selling of illegal drugs, but that doesn't stop a lot of people from trying. Again, we meet our old problem of supply and demand, and where there is demand for shark fin, there will be supply, legal or not.

And while it is true that Australian waters are visited by fishing boats from other countries, we are very good at killing our own sharks. Like I said earlier, I grew up eating flake, which is what we'd call shark meat. Despite the high level of toxic metals in the meat it is still for sale at the time of writing, and moreover, there's no requirement for the seller to test for toxin levels, label what shark species it came from or from where in the world. Let's say you were a consumer who wanted to eat shark. You don't know if what you're buying is from a less threatened species, or a more threatened one from a struggling shark population. Keep in mind that shark species in some parts of the world are close to being wiped out, but still survive in others. You might think what you are eating is from a sustainable species, but you actually just made your contribution to wiping out that animal from a certain part of the world. It's all a bit confusing, but the solution is extremely simple: let's just not eat sharks.

In Australian waters, some animals that meet the criteria for being endangered, such as the scalloped hammerhead, are instead given a conservation dependent status. This loophole allows them to continue to be employed for commercial use – in other words, killed – which is really the opposite of what you should be doing to an animal that's on its way to extinction.

This 'protected but not really' status exists for a lot of sharks in Australian waters. For instance, check out this section from the government's website:

> Porbeagle, shortfin mako and longfin mako sharks are listed as migratory species under the EPBC Act.
>
> Provided an operator is fishing in accordance with an accredited AFMA fisheries management plan, the operator may keep and trade these migratory sharks that are brought up dead, however, live sharks must be returned to the sea unharmed as per advice from the Department of the Environment and Energy.*

So you can keep and sell the animal, as long as it's dead when you pull it out of the water? That's a great help, thank you. How about just don't pull it out in the first place?

Like the other Western countries we have covered, Australia has sports fishing for sharks, but I don't think we need to go into that again. Let's look instead at what is a national disgrace:

* https://www.afma.gov.au/environment-and-research/protected-species/sharks

The Western Australian shark cull

This is the common name given to a government policy that was the stuff of nightmares for animals and the people who care about them. In 2014, following the deaths of seven people over three years, the government ordered baited drumlines around public beaches to catch and kill large sharks.

Before we go any further into this, let's just look at that figure again. Seven deaths by shark attack in Western Australia over a span of three years. Since 2000 there have been 17 – an average of about one a year. To give you a bit of comparison, in 2018 alone there were 17 coastal drowning deaths in Western Australia, and lifeguard services performed 481 rescues.* Are we going to announce a cull on water because of this?

Just as politicians sometimes start wars for their own ego, or to shore up votes, so the politicians of Western Australia declared war against sharks. There was no science or sense behind their policy. It was simply the creation of a problem through fear tactics, and then the age-old political tactic of offering a solution to a problem you've created.

Not everyone was fooled. In fact there were large demonstrations against the cull, and it received international condemnation, particularly from animal rights groups.

* https://www.mybeach.com.au/media-release/wa-coastal-drowning-deaths-up-by-31-on-last-year/

For a start, the practice is barbaric. Sharks are caught on the baited lines that are attached to a buoy. Many die from this stress. Those that are found alive and are over 10 ft are then killed by gunshot. Congratulations, government of Western Australia, you just took the reproductive mature animals out of the ecosystem.

Great whites are supposed to be protected by the federal government, but they granted an exemption to allow the state government to carry out their murder. Dolphins, whales, sea turtles, rays and all kinds and sizes of sharks have been documented as being killed by these drumlines. Thanks to pressure from animal rights groups – and locals who recognise that the ocean does not belong to humans – the use of drumlines has been sporadic, but they are often deployed when there has been a shark attack or large shark sighting near a beach, and they have killed hundreds of mature animals. Considering how long it takes these shark species to reach sexual maturity, these drumline killings will damage shark populations for generations, and for what? So people can go in the sea and feel a little safer, despite the fact they are ten times more likely to die of drowning than shark attack?

At the time of writing this book in 2021/22, Australia is still one of the only countries in the world that kills sharks in the name of preserving human life, and as a proud Aussie that makes me sick to my stomach. In Queensland, known all over the world as the place of the Great Barrier Reef, hundreds of sea creatures are killed on drumlines, many of them protected or endangered.

We mentioned at the beginning of the section why sharks, particularly the big ones, are vital for the health of the reef, but what does science have to do with any of this? Politicians in Queensland are worried about their careers in the here and now, and as such they must answer to the hysteria they have helped cause. Any politician with half an ounce of sense and decency would answer the outcrying voices after shark attacks in the following way:

> Shark attacks remain extremely rare, but we must recognise that when we enter the ocean – their home – we are running the tiny chance that we could have an encounter with a shark. Even with the work of our lifeguards you are more than ten times more likely to die of drowning than shark attack, but you are making a decision to take that risk when you go into the water. We will provide lifeguards. We will provide hospital care. We will provide drones and boat patrols in an attempt to give early warning, but we will not kill sharks, because we need sharks. The Great Barrier Reef is vital to Australia, and the world. If you don't want to run the tiny chance of ever being close to a shark in the water, then you don't have to. Stay out of the ocean, and this will never be an issue for you.

Will that ever happen? Of course not, but that is what a politician would say if they were speaking the truth. Instead, they continue with this mindless, barbaric policy of drumlines and nets. Far from keeping animals from the beach, they actually entangle them, and the struggles of these dying animals then draw in bigger predators to feed.

So what could be used instead? Well, there are several promising new technologies, including eco-friendly kelp barriers that use magnetic stimuli to dissuade sharks from passing through. Prototypes of these have been used successfully in the Bahamas and South Africa.* Drones and helicopters can use technology and good old-fashioned eyesight to spot big sharks. Obviously, the benefit of drones is that they are cheaper to use, and personally, I feel like helicopters make people feel as if they're in a scene from *Jaws*, and that just ups their desire for sharks to be taken out of the picture.

One of our biggest tools in the battle against this senseless slaughter is education. This comes in two forms: educating people about best practice, such as advising them to avoid swimming at dusk and dawn, and educating them about the importance of sharks so that they will not want to support shark culls. This slaughter is politically driven, and if politicians know it will cost them their job, they'll stop doing it. It's as simple as that.

Of course, the easiest way of avoiding shark attack is by not going in the water, and if you do go in, assume your own risk. Should we bulldoze Mount Everest to prevent another mountaineer from dying on its slopes? If someone suggested that we flatten every mountain to

* https://www.euronews.com/green/2021/03/10/why-kelp-forests-are-crucial-in-the-fight-against-climate-change

save the occasional mountaineering death, people would say, 'That's crazy. If they don't want the risk, they shouldn't go up there.' I agree, and the same applies to sharks and the ocean. Don't want that small bit of risk? Then stay out of the ocean. There are thousands of pools where you can go swimming instead.

Australia's attitude towards sharks is incredibly frustrating, particularly when we compare it with our next case study, but on the other side of the world may lie the key to turning Australian attitudes to sharks around ...

THE BAHAMAS

Before you start thinking that I included this tropical paradise just to justify a research trip to my accountant, these islands are in fact on the leading edge of shark conservation, and a very good case to celebrate and learn from.

Global FinPrint, the largest ever shark conservation study of its kind, has declared the Bahamas a 'world leader' in the field, which is no small praise. While shark populations around the world have been savaged by overfishing, they remain stable around these islands, and have done so for decade after decade.

The bottom line is that sharks in the Bahamas are protected, and as a result of this their populations have remained stable, the ecosystem has functioned as it should, and the island's shark inhabitants contribute to the economy by remaining alive, not dead. I've mentioned several times in the book that I've been part of documentary shoots in Caribbean waters, but you don't have to be a *Shark Week* presenter to get in among these awesome animals, and people flock from all over the world to see the sharks from boats, or even free dive and scuba with them. In the shark system of the Bahamas, everyone benefits.

Locals on the island confront sharks on a regular basis. As a small group of islands, fishing has always been part of the economy, and so has a relationship with the sea around them, including its inhabitants. The Bahamas have become a mecca for shark research, with several permanent institutes and many flying visits from scientists who come from all over the world.

We mentioned longlines earlier, which are long fishing lines with several baited hooks attached. While these are meant to catch fish like tuna, they bring in all kinds of bycatch, including a large number of sharks. The Bahamas banned their use in the 1990s, putting them literally decades ahead of most countries who still permit their use at the time of writing.

Then, in 2011 the Bahamas declared its exclusive economic zone – an area that covers 200 miles from the shoreline – a protected shark sanctuary. This protection banned shark fishing and prohibited the sale, trade and possession of shark or shark parts, and it does so over a 650,000-square-kilometre area.

So why do people in the Bahamas go along with this? Surely not everyone loves sharks as much as we do?

That's true, but what a lot of people do have in common is their love of money. We all have bills to pay, and I totally understand that a fisherman is going to put his kids ahead of a shark, but that's the great news – they don't have to. It is estimated that a killed reef shark is worth about $50 for meat. On the other hand, over its lifetime, that shark can bring in $250,000 worth of tourism to the local economy.* As further evidence to support this idea that sharks are worth more alive than dead, a study by the Cape Eleuthera Institute of 19,000 divers revealed that 43 per cent of these visitors came for the sharks.

We must also not underestimate the power that diving with a shark has in changing a person's relationship with an animal for the rest of their life. I challenge anyone to dive with these magnificent creatures and not come away from the experience as a shark rights advocate. Once you've seen them in their natural environment, and are witness to their true grace, and realise they are not the monsters you thought, then you will want to protect them, I promise you.

Governments, and citizens, of coastal countries should be looking at the Bahamas as an example of what is possible. Not everywhere has the golden beaches and warm waters of the Caribbean, but a lot of shark tourism is research and TV based. Look at the next nearest landmass, America. Florida specifically is only 50 miles from the Bahamas, and the state does already protect some animals, like the great hammerhead. Imagine if they decided to continue the amazing work of the Bahamas, and extended the protection to all sharks in their own waters. Come on Florida, you can do this!

* https://www.oceanographicmagazine.com/features/true-value-shark-sanctuaries/

I am so excited by what is happening in the Bahamas, and if you are one of the people who has made it happen – either through your research, your activism, or by putting your hard-earned money into the local economy there to visit sharks – then you have my eternal thanks. The oceans don't need to be a nightmare for our sharks, not at all. The Bahamas has shown that it can be a paradise, and one which humans and sharks share happily together. Hopefully, countries like Australia will take note, and they too can prosper from shark tourism and all the benefits it provides.

THE MALDIVES

The Maldives is an archipelago of islands in the Indian Ocean, and in 2009 it followed in the footsteps of the Bahamas and declared its first shark sanctuary. At the time of writing it continues to be one of the world's frontrunners in shark conservation, with an area of some 90,000 square kilometres of ocean giving full protection to sharks. This means no finning and no fishing.

Like the Bahamas, the Maldives has a strong tourist economy that accounts for a large part of the island's revenue. The Pew Environmental Group puts the revenue generated by the island's beaches and reefs at just over a quarter of the island's gross domestic product (GDP).*

* https://blueocean.net/maldives-creates-worlds-2nd-shark-sanctuary/

One of the reasons that this project has been successful is because the government of the Maldives didn't just ban shark fishing and tell the people in their industry 'good luck, you're on your own'. Rather, they actively helped them to find new careers. Like the former fishermen in the Philippines who protect whale sharks, who better to protect these animals than the people who have lived their lives on the sea? Shark conservation should be about encouraging people to change their ways through education and incentives rather than punishment, and that is exactly what we're seeing in these islands.

In 2021 rumours began to circulate that the government of the Maldives was planning on lifting the fishing ban. This was based on the ruminations of the Minister of Fisheries, Zaha Waheed. It prompted a meeting with shark conservationists, who reported that Waheed said that 'The Maldivian government is definitely not lifting the ban.' Apparently, it seems that some of the confusion came from the proposition of a new longline tuna fishery. Longlines inevitably catch sharks as bycatch, and given that they are protected, this of course raises legal issues. There are Maldivian ministers who oppose the shark ban and sanctuary, so keeping it in place may well be a constant battle for generations to come. We've seen in Australia how the government can find loopholes to allow the killing of sharks, even when they're said to be protected, so the situation in the Maldives is one to keep a close eye on, but with plenty of reason to be optimistic.

As with so much, it really could be money that decides this, and so far sharks have favourable odds in that department. A study conducted in 2018 found that sharks were responsible for a 15 per cent increase in scuba tourism, and if shark populations declined it could cost the local economy about $24 million a year.* As the demand for shark fin declines, and more and more people realise that shark meat is something to avoid for their own health, the cost/benefit analysis will continue to swing further and further towards supporting shark conservation. Now, the big question is, what can you and I do to make that day come even sooner?

OUR OVERTON WINDOW

There is a theory in politics known as the Overton window, which is when a policy becomes politically acceptable and mainstream to a population at a certain time. The American policy analyst Joseph Overton stated that an idea's political viability – in our case, the protection of sharks – depends on whether it falls in the range of mainstream opinion rather than a politician's individual preferences.

When an idea is new, it is often unthinkable. Can you imagine if you'd brought up the idea of sharks having rights fifty years ago? You'd have been laughed out of the room or thrown into a lunatic asylum. Let's remember that in the not-too-distant past, women didn't get to vote, gay people did not have the right to marry and there was segregation by race in some societies. At first

* https://www.sciencedirect.com/science/article/pii/S0261517718301201

the people who said we should have equal rights were branded as radical, but as more people became convinced of it this radical idea began to be regarded as acceptable, and then as sensible. At this point, according to Overton, it becomes a 'desired idea'. People want to share this view, and convince others of it, because they believe it is *right*. When enough people believe firmly in an idea it becomes popular, and what is popular often becomes policy.

Without doubt, I believe that in terms of shark conservation we are at the sensible stage of the Overton window. If you explain to people how many sharks are killed, and why that is bad, very few of them will not think that protecting these animals isn't sensible. The more the general public learns the facts, the sooner we can make the idea of shark conservation popular, and then we can make it policy – so written into law, and have those laws be rigorously enforced.

As evidenced by the fact that there is still inequality in society, even when an idea becomes desired it can take a long time to become policy, and for that policy to make a material and meaningful difference. Unfortunately, we don't have time to spare with sharks. With so many of them vulnerable to extinction, we need to take the Overton window and strap a rocket to it, and the way we do this is through personal action and collective activism.

INDIVIDUAL ACTION

With so much stacked against sharks, what difference can an individual make? The honest answer, if it was you and you alone, is not a lot. However, when we combine your actions with my actions, and another person's, and another person's, we start to become far more powerful by the sum of our parts.

Hopefully, for your own health as well as the shark's, you've been put off eating shark meat again because of the dangerous metal content, but I'm fully understanding of someone who still wants to eat fish. I get it. When something has been in your life for so long, it's impossible to think of life without it, but that's the problem – if we don't stop eating fish, we will be without them, and forever. Once that evolutionary chain is broken, then it's gone. It took sharks 450 million years to get to this point. Do we really want to wipe them out in a century because we're not willing to adjust our diet?

The fishing industry is all based on supply and demand. If just 10 per cent of people said, 'I'm done with eating fish,' just think what an impact that would have on the industry. It could be the difference between these large fishing companies being in profit, or out of profit, and if there's no profit to be made then they will shut down. Of course, the counter to this argument is that if there's no profit to be made then companies will just look for cheaper ways to fish, and these are often even more detrimental to wildlife, and that's why the next thing we need to do as individuals is just as important.

We must all become our own advocate and activist for the oceans.

Every single one of us has a voice. We talk to friends, family and strangers. We interact on social media, in our social life and at our place of work. Every year we have thousands upon thousands of interactions. Just think about that, and what could happen if you used some of those interactions to talk about why we need to act now, before it's too late. If you convinced five people to stop eating fish one year, and each of them convinced five other people, and so on, and so on, then your actions keep on multiplying even without you being around. That is the power of communication and community, and when action meets opportunity we solve a need. In this case, the need is to solve the crisis in our oceans, and the opportunities to do that are all around us. We don't need to go looking for them – step one is to just say, 'I'm not going to be a part of this.' If every one of us did that today, the crisis in the oceans would be solved in seconds. Of course, I'm not naive enough to think that will happen, but I do think there is a case for optimism.

ACTIVISM AND SHARK PROTECTION/ CONSERVATION GROUPS

One of the best parts of my day is when people reach out to me about sharks. Sometimes people want to talk shark facts, and others have questions about my shark-attack experience, which I then try and direct into more shark facts. I also get people asking me how they can get into doing what I do on TV, and I'm probably not a good presenter to ask, because my career definitely came about by accident after losing half my arm and leg, and I don't really recommend that. So, unless you have an uncontrollable urge to be almost eaten alive, be on the edge of death during surgical operations, lose the first career that you loved, lose riding the motorcycle that you loved, have people stare at you wherever you go, and live in constant pain and discomfort, then I would definitely not try the De Gelder method of auditioning for *Shark Week*.

Now that we've covered that, let's get to my favourite question that people ask me: 'How can I help protect sharks?'

There's actually a lot we can do as individuals in this area, but true power comes when we work together, and thankfully there are some incredible organisations that I can direct people towards.

There are so many wonderful humans and groups working on the frontlines to save the shark's realm, and just like our shark species, they come in all shapes and sizes.

As we've discussed, there are a lot of people who aren't just indifferent to the massacre of sharks, but they actively participate in it, or encourage it. Most governments are involved in this wrongdoing, and while people will say that what they are doing is legal, I would say that it is immoral, and anyone being honest will tell you it's not sustainable.

When you're dealing with these bureaucratic leviathans, not to mention the societal opinions entrenched through myth and popular culture, you have to take a multi-pronged approach to change people's core beliefs – which are our most deeply held assumptions about ourselves, or in this case, the world.

Let me tell you from experience that, as someone who used to eat meat every day of his life, it was an incredibly painful experience for me when I realised that my core belief had been wrong. I had convinced myself that I needed meat to be powerful and strong, and that core belief was so deeply ingrained in me that I kept going back to it. Breaking away from a deeply held core belief is like walking away from a partner that you love. It's incredibly hard and distressing, and it's only when you've been clear of it for some time that you can look back and see objectively how toxic it was.

It's hard to confront the idea that how we've been living our lives has contributed to harm, particularly if that harm was avoidable, and if it's irreversible. Our brains are great at protecting us, and so when confronted with these truths we often become angry and defensive. We try and rationalise what could not be rationalised by an objective person, but when holding on to a core belief we are anything but objective, because we feel like our entire being is under attack. It's not just the idea that someone is challenging us – it feels like they are challenging our soul.

This is called cognitive dissonance, and it is one of the greatest obstacles we face when trying to change people's perceptions on sharks, shark fishing and the important role sharks play in ecosystems. If you, the reader, understand that these issues are a problem then you probably can't understand why someone would defend them, but you must realise that the person is not necessarily defending the practices; rather, they are defending themselves for having participated in the slaughter.

Marketing slogans like 'line-caught tuna' make people believe there's no harm being done to other species. Most people really have no idea about what's going on, or the scale of it, and then of course there are other people who just don't care. The truth is that all of this information is just an internet search away, but that can be said about so many things. The fact is that, unless someone plants an idea in a person's head to look, then they never will. That is why it's up to us, as people who love sharks, to be the voice for those who can't speak for themselves, and to help guide people in the direction of the truth.

Keep in mind that everyone is going through something in their life, and that just because sharks aren't their priority, doesn't mean they don't care at all. Some people will also be receptive to new ideas at different points in their life. In my personal experience, people with entrenched opinions come in all shapes and sizes, but I've been surprised at how people I didn't ever think would change their mind eventually do so, and passionately. Because of the power of cognitive dissonance, it is very unlikely that you will change a person's mind by battering them over the head with facts that back up how bad their choices have been. Instead, you must lay a breadcrumb trail for them to follow and let them come to their conclusion in their own way. If a person feels that they have made the discovery and revelation, rather than being shamed into it, then they are far more likely to stick with that new way of thinking. Besides, we don't want people to be miserable and support sharks and our oceans because they feel bullied into it. We want them to do it because they realised that sharks are awesome, and we all benefit from healthy oceans. Only together can we save our sharks.

Now, if you're looking for groups at the tip of the spear that you can support, I have listed some at the back of this book. They make a big difference for our sharks, and many are run by mates of mine. They are hardworking, honest people who put their heart and soul into what they do, and any of them would be deeply grateful for any support you can give them. Donating time helps as much if not more than donating money, and even doing something as simple as sharing their work on social media, or signing and sharing their petitions to governments, can make a big difference. All of these things add up and help to create a wave that changes culture. We can make a change, but we need to act soon, and decisively. Time is running out, and when our sharks are gone, nothing in the world can bring them back.

CONCLUSION

> 'I'm not saying that everyone should swim with sharks, but sometimes you have to jump over your own shadow in order to learn something that you will never forget for the rest of your life. Then you know you can conquer your fears.'
>
> *Heidi Klum*

We all have our fears: heights, spiders, enclosed spaces, there are even some of us who fear, literally, *everything* (it's called panphobia). These fears often stem rationally from circumstances we've encountered during our lives, while others come from irrational programming placed into our psyche by external forces – such as movies and media in the case of sharks. Chances are that if you've read this book, you haven't given in to such hysteria and share my love and fascination for our finned friends. In my case, I have chosen to leverage my story of overcoming my fear of sharks, even after almost meeting my end in the jaws of one, as an example for others in the hope that people will see and understand that if a shark-attack survivor can rise above this fear through knowledge and compassion, then anyone can.

I didn't agree to write this book because I think I know everything there is to know about sharks, or because I think my story is worth more than anyone else's. I wrote it because I know there are people out there, just like me now, who hold a deep respect for sharks, and because I know there are people out there who are like I used to be, believing we'd be better off killing them all. I wrote it because I want to share my appreciation with anyone that will listen – and especially with those that fear them – of how truly amazing these animals are. Sharks are an apex predator, yes, but we humans are the absolute top of the food chain and we consume everything in our path. We are often irresponsible and reckless with the world around us. Just taking a stroll along any coastline on the planet, you can see in the sand our destructive force through the trash that endlessly washes up on the shore. Pretty little plastic pieces that mother birds feed their babies. Plastic bags that look like jellyfish and clog up the stomachs of turtles and whales. Micro-plastics that are consumed by phytoplankton, which then end up inside small fish, then bigger fish, and then sharks, and even us. The run-off from pesticides, herbicides, fungicides and cattle leach into our water tables and rivers, polluting waterways and creating dead zones in our oceans. Our human reach has no limit, and if we cannot see and correct the error of our ways, then our species is in line for its own disaster. As destructive as our capability to impact the planet can be, it is also our superpower, and just as incredible as a shark's ampullae of Lorenzini.

We don't have to continue along this path. We can influence the direction of our existence through words and actions. They may start small, but just like the snowball in the old *Looney Tunes Cartoons*, with each revolution it grows more and more powerful. With every word we speak and with every heart we touch in defence and joy of the world around us, our snowball of compassion and understanding grows more powerful. So speak up. Share the knowledge and fire in your heart so that they might spread to others and enrich their lives too, and in turn help heal our world.

APPENDIX A

FURTHER INFORMATION

All you need to do to get your shark fix these days is to turn on the TV or computer and head to your favourite streaming platform.

There are a plethora of incredible humans that work tirelessly to bring you the shark education you need in the form of documentaries. As amazing as these animals are, many of these subjects we've already touched make for hard viewing, such as overfishing, finning, shark fishing tournaments and extinctions. Sharks have been having a rough time over the last 60 years thanks to us humans and our innovative ways to kill everything in our path, but some people are fighting back to protect our oceans. All of the people that are instrumental in creating shark- and ocean-based documentaries know who they are, and they don't do it for credit, but we salute you and thank you all the same. Creating documentaries is a team effort, and so is conservation. Below is a list – by no means exhaustive – of some further viewing on the subjects we have touched on in this book.

Rob Stewart: *Sharkwater Revolution*, *Sharkwater Extinction*
Shawn Heinrichs: *Racing Extinction*, *Mission Blue*, *Sharkwater Revolution*
Ali Tabrizi: *Seaspiracy*
Louie Psihoyos: *Racing Extinction*, *The Cove*, *Game Changers*
Philip Waller: *Extinction Soup*
Eli Roth: *Fin*
Madison Stewart: *Shark Girl*, *Sea of Life*, *Great Barrier Reef: The Next Generation*, *Envoy: Shark Cull*
Ocean Preservation Society: *Racing Extinction*, *The Cove*
Mark Monroe: *The Cove*, *Mission Blue*

APPENDIX B

SHARK CONSERVATION GROUPS

Knowing a problem is one thing, and making a change is another. Fortunately for us, there are groups who work tirelessly on behalf of sharks and our oceans. Many of the organisations listed below are run by friends of mine, and they will be grateful for any help you can give them. These people are true heroes in my eyes, and we're lucky to have them.

(Note: the descriptions below are from the organisations themselves. Please check out their websites for more information.)

Shark Allies: www.sharkallies.org

We are dedicated to the protection and conservation of sharks and rays. Our focus is on taking action, on raising awareness and guiding initiatives that reduce the destructive overfishing of sharks on a global scale. Shark Allies is a 501 (c)(3) non-profit organisation originally established in 2007 in Hawaii and incorporated in California in 2014.

Project Hiu: www.projecthiu.com

The goal of Project Hiu is to improve conditions above and below the surface, and enforce the idea that one person, and one shark fisherman, can make a difference.

Shark Angels: www.sharkangels.org

Globally connected, Shark Angels around the world are taking action locally, fuelled by empowering tools, a collaborative community and a shared passion. Through positive education, media and grassroots outreach, Shark Angels are changing the future for sharks. We are the next generation of shark conservationists, working independently and as a network of angels to bring about change – because sharks need all the guardian angels they can get.

Shawn Heinrichs: www.bluespheremedia.com

Shawn Heinrichs is an artist and Emmy-award-winning cinematographer, photographer and marine conservationist. His stunning and often stark artwork is fuelled by his passion to protect the oceans, and the profound recognition that people only protect what they love. His images have transformed communities, and he hopes his work will inspire people to act before it is too late. In recognition of Shawn's hands-on approach to marine conservation, he was presented with the 2011 Oris Sea Hero of the Year award.

Sea Shepherd: www.seashepherdglobal.org

Sea Shepherd is an international, non-profit marine conservation organisation that engages in direct action campaigns to defend wildlife and conserve and protect the world's oceans from illegal exploitation and environmental destruction.

Sharks 4 Kids: www.sharks4kids.com

The goal of Sharks 4 Kids is to create a new generation of shark advocates through access to a dynamic range of educational materials and experiences. Curriculum, games and activities will allow teachers to integrate shark education into their science programmes on an introductory, intermediate or advanced level. Students can access games, activities and info sheets to satisfy their own curiosity about sharks. Photos and videos from scientists and conservationists bring an exciting element into the classroom and show students the beauty of the ocean.

Shark Conservation Fund: www.sharkconservationfund.org

The Shark Conservation Fund (SCF) is a collaboration of philanthropists dedicated to solving the global shark and ray crisis. Our goal is to halt the overexploitation of the world's sharks and rays, prevent extinctions, reverse population declines and restore imperilled species through strategic and catalytic grantmaking.

Minorities in Shark Sciences (MISS): www.misselasmo.org

We were founded by four Black female shark researchers. We strive to be seen and take up space in a discipline which has been largely inaccessible for women like us. We strive to be positive role models for the next generation. We seek to promote diversity and inclusion in shark science and encourage women of colour to push through barriers and contribute knowledge in marine science. Finally, we hope to topple the system that has historically excluded women like us and create an equitable path to shark science. We believe diversity in scientists creates diversity in thought, which leads to innovation.

ACKNOWLEDGEMENTS

There are a great many people I'd like to thank for being instrumental in my journey with sharks.

First and foremost, my four Navy Clearance Diving teamates – Patto, Darty, Lauchie and Thommo – who kept me alive on that fateful day back in 2009. Without their calm under pressure and high level of training, I would not be here to share this book with you.

Thanks also to *60 Minutes Australia*, Peter Overton, David Alrich and Andy Taylor, and to Fiji shark diver extraordinaire Brandon Paige for taking me on my very first dive with bull sharks so that I could face my fear and see these animals in their true form.

A huge thank you to Discovery Channel for granting me the opportunity to be a small part of the phenomenon that is *Shark Week*. You don't always get it exactly right, but the world would be far less informed and in awe of our sharky friends if it were not for your steadfast dedication to creating the best shark content on the planet (even though I'm pretty sure you've tried to kill me a couple of times).

To all of the scientists, conservationists, cinematographers, camera operators, sound engineers, producers, directors, editors, runners and others whom I've worked with over the years – there are far too many of you to name but none of this would have been possible had it not been for your extremely hard work under immense pressure involving days, weeks and months dedicated to a single goal, creating amazing television that we can all be proud of.

A big thank you to all of the brave celebrities I've had the pleasure of introducing to the shark world over the years – Ronda Rousey, Will Smith, Mike Tyson, Rob Riggle, Damon Wayans Jr, Adam Devine, Anthony Anderson, Joel McHale, and David Dobrik. You put your trust in me to teach you and keep you safe. Hopefully I was able to give you an incredible experience that you'll never forget and will use to teach your friends, fans and followers in turn.

To my good mate and brother in arms, Geraint Jones. Thanks for coming on board with this project and helping me turn my jumble of stories, anecdotes and shark science into a story we can both be proud of. It was a tight timeline to get this sometimes very complicated piece of work together, but we got it done.

I accepted the task of writing this book not because I think I know everything about sharks or that I consider myself an expert. I did it solely because I've had some of the most amazing experiences of my life with sharks, and have been blessed to be allowed to share their home and space with them. I used the knowledge and personal experiences I built up over the years with them as a base to create a platform that I can share with all of you so you might gain a deeper insight into this beautiful world.

I don't have access to much of the new research or the latest publications in the world of shark science kept by research institutes and universities, but I've done my best to be as accurate as possible. A big thank you to Dr Craig O'Connell from O'Seas Conservation Foundation for agreeing to proofread and help me with corrections. This is a work of passion, not a science text book, but with the help of Craig I think we've done a pretty great job.

Finally, thank you to HarperCollins UK for reaching out and giving me the chance to share one of my favourite topics with the world.

INDEX

INDEX